MW01639669

Energy and Environmental Law and Policy Series

VOLUME 33

Editor

Introduction

Environmental protection and energy efficiency/security are important societal challenges. In order to tackle them, policy and legal frameworks are developed at national, regional and global level. Through study and best practices development, the challenges will prove to be solvable.

Contents/Subjects

Environment/Nature/Energy/Climate.

Objective

The aim of this series is to publish works of excellent quality that focus on the study of energy and environmental policy. Through this series the editors:

- contribute to the improvement of the quality of energy/environmental law and policy in general and environmental quality and energy efficiency in particular; increase the access to environmental and energy information for academics, non-governmental organizations, government institutions, and business; and
- facilitate cooperation between academic and non-academic communities in the field of energy and environmental law and policy throughout the world.

Readership

Academics and practitioners in environmental and energy matters.

The titles published in this series are listed at the end of this volume.

The Development of a Comprehensive Legal Framework for the Promotion of Offshore Wind Power

The Lessons from Europe and Pacific Asia

Edited by

Anton Ming-Zhi Gao
Chien-Te Fan

Published by:
Kluwer Law International B.V.
PO Box 316
2400 AH Alphen aan den Rijn
The Netherlands
Website: www.wolterskluwerlr.com

Sold and distributed in North, Central and South America by:
Wolters Kluwer Legal & Regulatory U.S.
7201 McKinney Circle
Frederick, MD 21704
United States of America
Email: customer.service@wolterskluwer.com

Sold and distributed in all other countries by:
Quadrant
Rockwood House
Haywards Heath
West Sussex
RH16 3DH
United Kingdom
Email: international-customerservice@wolterskluwer.com

Printed on acid-free paper.

ISBN 978-90-411-8397-2

e-Book: ISBN 978-90-411-8398-9
web-PDF: ISBN 978-90-411-8399-6

Printed in the United Kingdom.

Editors

Anton Ming-Zhi Gao is Associate Professor of Law at Institute of Law for Science and Technology, National Tsing Hua University (NTHU). He received his master degree from National Taiwan University. In October 2005, he began a Ph.D. program at K.U. Leuven, and received his Ph.D. in Law in December 2009. He currently teaches several courses, including Energy Law and Policy and European Environmental Law. He is the Taiwanese scholar with the greatest number of articles published in international energy and environmental law journals. Over the past number of years, he has been very active in international publications concerning investigations into the legal issues of international, European, Asian, and Taiwanese energy laws.

Chien-Te Fan is Professor of Law at National Tsing Hua University (NTHU), where he specializes in energy and environmental law, and the areas of biotechnology and ethic related issues. Prof. FAN received his LLB degree from Soochow University, LLM degree from University of Washington and JD Degree from Puget Sound University. He is also an experienced scholar involving in UNFCCC negotiation process since 2001 in Marrakesh. Moreover, as a speaker, he was also invited to join the side event several times that held in the COPs.

Contributors

Jeremy Firestone has a PhD in Public Policy Analysis from University of North Carolina and a JD from University of Michigan. He is a Professor of Marine Policy and of Legal Studies and is Director of the interdisciplinary Center for Carbon-free Power Integration at the University of Delaware. His research focuses on wind power, with emphasis on spatial planning, social acceptance, economics, and governance. He teaches courses on Renewable Energy Law, Climate Change Policy, Offshore Wind Power, and Ocean and Coastal Law. He has published in *Proceedings of the National Academy of Sciences, Energy Policy, Wind Energy, Renewable Energy, Journal of Environmental Planning & Management, and Coastal Management* and is on the editorial board of the journal *Energy Research and Social Science.*

Sandra Cassotta is Associate Professor in International, Environmental and Energy Law at the Department of Law of Aalborg University (Denmark) and teaches International Environmental Law, Climate Change, Climate Change and Migration Law, Arctic Environmental Law and Geopolitics. She specializes in environmental damage and liability problems in a multi-level context. Included in her area of interests are human rights, law of the sea (UNCLOS), and environmental security. She is non-resident Research Fellow at the Institute for Security and Development Policy in Stockholm (Sweden) working under the Sino-Nordic Arctic Policy Program. She is also jump external lecturer at Leuven University KU on Malta at the University of Malta (La Valletta) and adjunct professor at the International Centre of Ocean Governance (ICOG) at Western Sydney University, School of Law (Australia). She can be reached at sac@law.aau.dk, Aalborg Universitet, Juridisk Institut, Niels Jernes Vej 6B, 2.05b, 9220 Aalborg Øst, Denmark.

Ulla Steen is associate professor at the Department of Law of Aalborg University. She is Head of the Department of Law. Her areas of interest are Environmental and Energy Law, Environmental Impact Assessment, and EU Law. She has several years of practical experience both in the private and public sector and has a strong international working experience even outside academia. She was associated attorney of law and worked for NIRAS, several years. Her research focus is to improve sustainable development worldwide by combining business strategies with environmental, EU and Energy Law research related fields from different angles and its relation to climate change and energy global processes. She can be reached at: usteen@law.aau.dk, law.aau.dk, Aalborg University, Niels Jernes Vej 8a, DK-9220 Aalborg.

Navraj Singh Ghaleigh is Senior Lecturer in Climate Law at the University of Edinburgh where he directs two LLM programmes: *Global Environment and Climate Change*; and *Law and Chinese*. He publishes widely in the field, has advised various Parliamentary and Executive bodies, and is a board member of both the *Society of Legal Scholars* and *Climate Strategies*.

Dr. Haifeng Deng is the associate dean and tenured associate professor of Law School, Tsinghua University, China. He is also the vice director of the Center for Environmental, Natural Resources & Energy Law of Tsinghua University and the senior research fellow of the CDM Development and Research Center of Tsinghua University. His research is focusing on the environmental law, nature resource law and energy law. At the same time, he also does some public service to the legal research societies. Now, he is the standing director of China environmental law research society, the director of environmental law research society of Beijing law science society, the standing director and vice general secretary of environmental law research society of China environmental science society and the formal expert of the official version of the legislation on Chinese Climate Change Arrangement Law.

Dr. Alex Wawryk received First Class degrees in Economics and Law, and a PhD in Law, from the University of Adelaide. She is Senior Lecturer in Contract Law and Mining and Energy Law at the Adelaide Law School and is a Barrister and Solicitor of the Supreme Court of South Australia. Alex conducts research into various aspects of energy and natural resources law, including renewable energy law, mining and petroleum regulation, and environmental law. She is an associate editor for OGEL, a specialist online database for Oil, Gas and Energy Law. Alex is a Board member of the Environmental Defenders Office (SA) Inc, and is a member of the IUCN Specialist Group on Energy Law, the Australian Resources and Energy Law Association, and the Institute for Mining, Energy and Resources at the University of Adelaide.

Eubong Lee got a PhD in Law in Seoul National University, South Korea. She has researched for national legislative policy theoretically as well as practically in Korea Legislative Research Institute, a national policy think tank. Her special focus on research is environmental law, renewable energy law and public ethics law. Also, she has taught environmental law in the Graduate School of Environmental Studies, Seoul National University.

Thomas Schomerus (University Professor Dr. iur. Dr. h.c.) studied Law at the Universities of Hamburg and Göttingen from 1976–1981. After finishing his legal traineeship with the second state exam in 1988, he started his career as an administrative official for the city of Hamburg. In 1996, he was appointed professor at the former University of Applied Sciences, Lüneburg. Since 2004, he holds the chair of public law, in particular energy and environmental law, at Lüneburg University. He was awarded an honorary doctor's degree by the Georgian Technical University, Tbilisi, in 2005. Schomerus worked on several international and national research projects and has published broadly in the fields of energy and environmental law. In 2014, he was appointed Judge at the Higher Administrative Court of Lower Saxony.

Christian Maly holds an MA in Management and Business Development and an LLB from Leuphana University of Lüneburg, Germany. He is specialized in environmental and renewable energy law with a special focus on wind energy. Christian Maly worked as a research associate and is currently a PhD candidate and lecturer at Leuphana University of Lüneburg.

Dr. Catherine Banet is associate professor at the University of Oslo, Scandinavian Institute of Maritime Law, Department of Petroleum and Energy Law, Norway. Her core fields of legal expertise include energy law, petroleum law, environmental law, competition law (in particular state aids), EU/EEA law and law of the sea. She has background from private law practice (Norway, France), the European Commission (DG ENV), US diplomatic mission and academia. She is one of the four academic coordinators of the professional LLM programme North Sea Energy Law Programme (NSELP) and co-organizer of the annual European Energy Law Seminar (EELS). She is a member of the Academic Advisory Group of the Section on Energy, Environment and Infrastructure Law of the International Bar Association, board member of the ECOHZ Renewable Energy Foundation (Norway) and member of the committee on energy and environmental law of the Norwegian Center for Continuing Legal Education.

Summary of Contents

Table of Contents

Preface

After the golden age of onshore renewable energy over the past few decades, several factors have led to the recent movement toward offshore renewable energy. These factors include existing use of strong-wind sites, lack of space and protests against further development. Of all the offshore renewables, offshore wind has shown the strongest growth over the past few years. For this reason, it could contribute most effectively to our energy future.

This book can be seen as a continuation and update of the previous book entitled *Legal Systems and Wind Energy: A Comparative Perspective* (edited by: Helle Tegner Anker, Birgitte Egelund Olsen, Anita Rønne, by Kluwer, 2008). This book aims to apply the previous book's analytical structure in the Pacific Asia Region (the US, Australia, China, Korea and Taiwan) and in other parts of Europe including Germany, France, Italy, Norway, and the UK. Additionally, this book will focus solely on offshore wind power. It will focus on the scheme for wind turbines, but also on the key infrastructure of grid, construction harbor, and vessels. Finally, for the regulatory section, this book will address both the environmental, health and safety (EHS) concerns related to offshore wind, and its incentives.

The book aims to provide a holistic approach to key lessons that we can learn from Germany and the UK, the leaders in offshore wind power development (Part I. Offshore Wind Power of Highly Developed Countries in Europe). It strives to determine if we could apply a similar legal structure in the Pacific Asia Region (Part III. Offshore Wind Power of Developing Countries in the Pacific Asia) and in other European countries (Part II. Offshore Wind Power of Developing Countries in Europe).

Usually, the public would have the perception that policy initiative and subsidy mainly drive the development of offshore wind power. Yet, from the lessons of Offshore Wind Power of Highly Developed Countries, the development of a strong and sound legal framework could also have the same importance. In this book, such a framework would embrace three key elements.

First, before moving into the details of legal design for offshore wind power development, a well-functional Institutional Design for Offshore Wind Power is fundamental to pave the way for furthering good incentives and regulations. Such competent authority plays the role not only of drafting good policy but also of avoiding red tape during the development stage of offshore wind power. From the surveyed countries, the one-stop-shop/integrated procedures models in Germany and the UK are highly recommended. In the rest of the countries, fragmented authorities and procedures may post tremendous procedural burden to offshore wind power, whose nature is more complex than the rest of renewable electricity installations.

Second, the main drivers of all renewables, including offshore wind, are the incentives and subsidy. So far, most attention has been paid to the key incentive scheme, such as feed-in tariff or tendering. However, a good mixture of main incentives - feed-in tariff (FIT), renewable portfolio standards (RPS), renewable obligation (RO), and tendering - and supplementary schemes - investment subsidy, low interest loan, taxation credit, favorable grid cost-sharing rules - is indispensable. Particularly, offshore wind power faces higher barriers and risk of loan and finance issues than onshore wind power. A balanced approach for structuring the incentives scheme to reflect the needs of offshore wind power industries is necessary. Of course, it is also related to the needs of national policy direction. Does the country at stake focus on the real deployment of offshore wind power to meet its emission reduction and renewable energy goals? On the other hand, do the goals also involve the development of local offshore wind power industries to create green jobs?

For most of renewable energy, money talks and incentives could play a key role. Yet, for offshore wind power, regulatory barriers pose a lot of threat to the development of offshore wind power projects. A license scheme stands in the center and is surrounded by the environmental health and safety (EHS) concerns of the projects. In this regard, how to develop an appropriate license scheme, and related EHS regulations, such as Environmental Impact Assessment (EIA), and Strategic Environmental Assessment (SEA) and Siting Related Planning, could contribute to the development of offshore wind power as well.

With a view to deliberate the legal issues above in a careful manner, on August 22–23, 2016, the Institute of Law for Science and Technology and the Bioethics and Law Center of the National Tsing Hua University, the Office of Energy Policy for Bridging and Communication, the National Energy Research Programme, and the College of Intellectual Property Studies of National Taiwan University of Science and Technology (NTUST) organized an International Conference on Comprehensive Legal Framework for the Development of Offshore Wind Power Around the World, with the funding support from the Ministry of Science and Technology, the Research Center for Humanities and Social Sciences, National Tsing Hua University, and the Office of Research & Development, National Tsing Hua University. Special thanks go to the Research Center for Humanities and Social Sciences, National Tsing Hua University, Taiwan, whose funding has made the combination of this international conference and book project possible.

Anton Ming-Zhi Gao
Associate Professor, Institute of Law for Science and Technology,
National Tsing Hua University
Fan Chien-Te
Director and Professor, Institute of Law for Science and Technology,
National Tsing Hua University

PART I Offshore Wind Power of Highly Developed Countries in Europe

CHAPTER 1

Legal Framework to Develop Offshore Wind Power in Germany

Thomas Schomerus & Christian Maly

§1.01 HISTORICAL REVIEW AND CURRENT SITUATION OF OFFSHORE WIND ENERGY IN GERMANY

Energy transition ("*Energiewende*") is one of Germany's big challenges. Aiming at a complete turnaround in energy policy, this megaproject targets the shift from using fossil fuels and nuclear energy[1] to renewable energies.[2] German energy transition is embedded in the current policy framework of the European Union (EU) Climate Change and Energy Package (20–20–20 Package) with three binding key targets for 2020: a 20% improvement in energy efficiency, a 20% cut in greenhouse gas emissions (from 1990 levels) and, a total of at least, 20% of the EU's total energy from renewable energy sources.[3] The Renewable Energy Directive of April 2009 establishes mandatory national targets as well as measures for each EU Member State to meet the renewable energy targets of the 20–20–20 Package.[4] According to this Directive, by 2020, 18% of energy consumption in Germany must be provided by renewable energy sources. The

1. Nuclear power plants will be phased-out by 2022, *see* sections 1 and 7 Act on the Peaceful Utilisation of Atomic Energy and the Protection against its Hazards (Atomic Energy Act), English translation available under http://www.bfs.de/SharedDocs/Downloads/BfS/EN/hns/a1-english/A1-07-16-AtG.pdf?__blob=publicationFile&v=10 (accessed November 26, 2016).
2. *Maly*, Legal aspects of local engagement: Land planning and citizens' financial participation in wind energy projects, in: Peeters & Schomerus, Renewable Energy Law in the EU – Legal Perspectives on Bottom-up Approaches (2014), 210 with further proofs.
3. *See* Council of the European Union, Energy and climate change – Elements of the final compromise, December 12, 2008 (17215/08), available under http://www.consilium.europa.eu/uedocs/cmsUpload/st17215.en08.pdf (accessed November 26, 2016).
4. Directive 2009/28/EC of April 23, 2009 on the promotion of the use of energy from renewable sources, Official Journal of the European Union L 140/16.

new policy framework for climate and energy for the period from 2020 to 2030 sets an EU-wide target for renewable energy sources to provide at least 27% of total energy consumption by 2030.[5]

To reach these targets, Germany must expand the use of renewable energy sources. According to the federal government's strategy, wind energy should become the key component of the German energy transition.[6] In recent years, Germany has experienced an upswing in the renewable energy industry, especially in the onshore wind-energy sector. In 2015, renewable energy already contributed 32.6% of gross electricity consumption in Germany.[7] By 2016, 26,561[8] onshore and 835[9] offshore wind turbines were installed. However, onshore wind energy projects face an increasing level of local resistance which may delay or even block new projects.[10] A study from 2015 shows that the majority of Germans support the expansion of renewable energy sources (93%),[11] but, with regard to specific local onshore wind-energy projects, support is significantly lower.[12] There are well-established approaches for overcoming challenges for local acceptance, such as financial participation models on a voluntary

5. The new policy framework is based on the EU Commission's proposal COM(2014) 15 final, A policy framework for climate and energy in the period from 2020 to 2030, available und er http://ec.europa.eu/smart-regulation/impact/ia_carried_out/docs/ia_2014/swd_2014_0015_en.pdf (accessed November 26, 2016).
6. *German Federal Government*, Energiekonzept für eine umweltschonende, zuverlässige und bezahlbare Energieversorgung (2010), 6, 9, available under https://www.bundesregierung.de/ContentArchiv/DE/Archiv17/_Anlagen/2012/02/energiekonzept-final.pdf?__blob=publicationFile&v=5 (accessed November 26, 2016).
7. *Federal Ministry for Economic Affairs and Energy* (BMWi), Erneuerbare Energien in Deutschland - Daten zur Entwicklung im Jahr 2015 (2016), 3, available under http://www.erneuerbare-energien.de/EE/Redaktion/DE/Downloads/erneuerbare-energien-in-zahlen-2015.pdf?__blob=publicationFile&v=4 (accessed November 26, 2016).
8. *Deutsche WindGuard*, Status des Windenergieausbaus an Land in Deutschland - 1. Halbjahr 2016 (2016), 1, available under https://www.wind-energie.de/sites/default/files/download/publication/factsheet-status-des-offshore-windenergieausbaus-deutschland-1-halbjahr-2016/factsheet_status_windenergieausbau_an_land_1._halbjahr_2016.pdf (accessed November 26, 2016).
9. *Deutsche WindGuard*, Status des Offshore-Windenergieausbaus in Deutschland - 1. Halbjahr 2016 (2016), 1 with numbers for OWFs feeding electricity into the grid, available under http://www.windguard.de/_Resources/Persistent/73e9b20264513bfff07ee44b637997d8b36733e6/Factsheet-Status-Offshore-Windenergieausbau-Halbjahr-2016.pdf (accessed November 26, 2016).
10. *Maly*, Legal aspects of local engagement: Land planning and citizens' financial participation in wind energy projects, in: Peeters & Schomerus, Renewable Energy Law in the EU - Legal Perspectives on Bottom-up Approaches (2014), 211-213, with further proofs.
11. *Agentur für Erneuerbare Energien*, Akzeptanz für Erneuerbare weiterhin hoch, 27 Renews Kompakt (2015), available under https://www.unendlich-viel-energie.de/media/file/416.AEE_RenewsKompakt_Akzeptanzumfrage2015.pdf (accessed November 26, 2016).
12. *Bidwell*, The role of values in public beliefs and attitudes towards commercial wind energy, 58 Energy Policy, 190 (2013); *Wunderlich*, Akzeptanz und Bürgerbeteiligung für Erneuerbare Energien, Agentur für Erneuerbare Energien e. V.: Renews Spezial Akzeptanz & Bürgerbeteiligung für Erneuerbare Energien, No. 60, 10 (2012); *Wüstenhangen* et al., Social acceptance of renewable energy innovation: An introduction to the concept, 35 Energy Policy (2007).

or mandatory level.[13] Due to the rise in number of onshore wind-energy projects, suitable locations are limited and already often commercially exploited.[14]

The German Federal Government consequently supports the development of offshore wind energy,[15] backed by the European Commission.[16] The idea of using and expanding offshore wind energy is not new in German energy policy. The Marine Facilities Ordinance ("*Seeanlagenverordnung – SeeAnlV*")[17] of 1997 introduced a legal framework for planning and licensing Offshore Wind Farms (OWFs) and designated the Federal Maritime and Hydrographic Agency ("*Bundesamt für Seeschifffahrt und Hydrographie – BSH*")[18] as the responsible authority for granting approvals to OWFs in the German Exclusive Economic Zone (EEZ). From 2017 on, the new "Offshore Wind Sea Act" ("*Windenergie-auf-See-Gesetz – WindSeeG*")[19] will regulate planning, tenders for designated areas and licensing procedures for OWFs. Within the framework of this Act, the Federal Network Agency ("*Bundesnetzagentur – BNetzA*")[20] and the BSH are the most important authorities for the development and regulation of offshore wind energy.

German legislation has taken into account that financial promotion of renewable energies is of the utmost importance for its success. The Electricity Feed-In Act of 1990 ("*Stromeinspeisungsgesetz*")[21] introduced the first policy framework for supporting renewable energy sources. The 2000 Renewable Energy Sources Act ("*Erneuerbare-Energien-Gesetz – EEG*")[22] introduced a new dimension of renewable energy promotion by setting up a technology-dependent minimum price (feed-in tariffs) and the obligation for transmission operators to buy the electricity produced by renewable energy sources. Furthermore, a long-term offshore-wind-energy strategy was adopted by the

---

13. *Egelund Olsen*, Renewable energy: Public acceptance and citizens' financial participation, in: Farber & Peeters, Climate Change Law (2016), 476–499; *Maly & Meister*, Finanzielle Bürgerbeteiligung: Rechtsrahmen und Herausforderungen, in: Holstenkamp & Radtke, Handbuch Energiewende & Partizipation (2016) – forthcoming.
14. *Maly*, Legal aspects of local engagement: Land planning and citizens' financial participation in wind energy projects, in: Peeters & Schomerus, Renewable Energy Law in the EU – Legal Perspectives on Bottom-up Approaches (2014), 211 with further proofs.
15. *Reichardt* et al., Analyzing interdependencies between policy mixes and technological innovation systems: The case of offshore wind in Germany, Technological Forecasting & Social Change 106 (2016), 11–21; *Kostka & Anzinger*, Offshore Wind Power Expansion in German: Scale Patterns, and Causes of Time Delays and Cost Overruns, in: Kostka & Fiedler, Large infrastructure Projects in Germany (2016), 147–148; 155–156.
16. *European Commission* (EC), Offshore wind energy: Action needed to deliver on the energy policy objectives for 2020 and beyond (2008); *see*: http://eur-lex.europa.eu/resource.html?uri=cellar:13ed7fdb-684f-4e84-877a-95a3be850409.0004.03/DOC_2&format=PDF (accessed November 26, 2016).
17. Verordnung über Anlagen seewärts der Begrenzung des deutschen Küstenmeeres (Seeanlagenverordnung – SeeAnlV), January 23, 1997, Federal Law Gazette I p. 57.
18. *See* http://www.bsh.de/en/index.jsp (accessed November 26, 2016).
19. *See* Gesetzesbeschluss des Deutschen Bundestages, Bundesratsdrucksache 355/16 of 8 Juli 2016, available under https://www.clearingstelle-eeg.de/files/BR-Drs_355-16_160708_Grunddrucksache.pdf (accessed November 26, 2016).
20. *See* http://www.bundesnetzagentur.de/cln_1421/EN/Home/home_node.html;jsessionid=2938FA8E523FE27F160A9C64C71B26DE (accessed November 26, 2016).
21. Gesetz über die Einspeisung vom Strom aus erneuerbaren Energien in das öffentliche Netz, December 7, 1990, Federal Law Gazette I p. 2633.
22. Gesetz über den Vorrang Erneuerbarer Energien, March 29, 2000 (Federal Law Gazette I p. 305).

German government in 2002, setting up a target of 25 GW offshore wind energy-capacity by 2030.[23]

The EEG was amended several times[24] in order to adjust legal (e.g., the introduction of different feed-in tariffs for onshore and offshore wind energy by the EEG 2004, or the increase of feed-in tariffs for offshore wind energy by the EEG 2009) and technical requirements, taking the latest developments in the renewable energy sector into account and setting up incentives to improve technological innovation and cost efficiency. To overcome difficulties in providing grid connections under the regime of the EEG, the German Energy Industry Act (EnWG)[25] was amended in 2006. It transferred the responsibility of providing grid connections from project developers to the transmission system operator (TSO).

Moreover, in 2005, the Federal Government started fostering offshore wind energy development by establishing the "German Offshore Wind Energy Foundation," an independent institution targeting the promotion of the use of wind energy at sea by serving as a communication platform for stakeholders.[26] The wind farm "Alpha Ventus" was used for research projects and was supported financially by the Federal Ministry for the Environment, Nature Conservation, Building and Nuclear Safety ("*Bundesministerium für Umwelt, Naturschutz, Bau und Reaktorsicherheit – BMUB.*").[27] In 2010, the Federal Government introduced a new energy concept, confirming the offshore expansion target of 25 GW by 2030.[28] As an answer to financing problems regarding the grid connection for OWFs and the impacts of the financial crisis, it introduced the state owned KfW-bank loan program "Offshore Windenergie" as part of the "*10-Punkte-Sofortprogramm zum Energiekonzept der Bundesregierung.*" The program was designed to grant loans (EUR 5 billion) at market conditions for up to ten OWFs.[29]

In 2012, in order to address technical problems with the grid connection of OWFs, the Federal Government implemented a system change in the EnWG. From then on, the TSO and the project developer must negotiate the date for the OWF's grid connection.[30]

---

23. *German Federal Government*, Strategie der Bundesregierung zur Windenergienutzung auf *See* (2002), 7, available under http://www.loy-energie.de/download/Bundesregierung,%20windenergie_strategie_br,%2001-2002.pdf (accessed November 26, 2016).
24. EEG 2004, EEG 2009, EEG 2012, EEG 2014, EEG 2017.
25. Gesetz über die Elektrizitäts- und Gasversorgung (Energiewirtschaftsgesetz – EnWG) of July 7, 2005, Federal Law Gazette I p. 1970.
26. *Foundation Offshore Wind Energy* (2016), *see*: https://www.offshore-stiftung.de/en/mission (accessed June 28, 2016).
27. *Reichardt* et al., Analyzing interdependencies between policy mixes and technological innovation systems: The case of offshore wind in Germany, Technological Forecasting & Social Change 106 (2016), 16.
28. *Bundesregierung*, Energiekonzept (2010), *see* https://www.bundesregierung.de/ContentArchiv/DE/Archiv17/_Anlagen/2012/02/energiekonzept-final.pdf?__blob=publicationFile&v=5 (accessed June 28, 2016).
29. http://www.offshore-stiftung.com/60005/Uploaded/BvBuelow_R_090611.pdf (accessed June 28, 2016).
30. *Reichardt* et al., Analyzing interdependencies between policy mixes and technological innovation systems: The case of offshore wind in Germany, Technological Forecasting & Social Change 106 (2016), 17.

The SeeAnlV was also recast, introducing a plan approval procedure for the construction and operation of OWFs to be carried out by the BSH.[31] In recent years, the BSH specified relevant criteria for documents and studies required in the licensing process, e.g., the "Standard – Investigation of the Impacts of Offshore Wind Turbines on the Marine Environment (StUK4)."[32] Since 2013, in the case of connection problems to the grid, the TSO must pay 90% compensation of forgone revenue to the project developer, resulting in an offshore electricity surcharge for the final electricity consumer ("*Offshore-Haftungsumlage,*" up to 0.25 cent/kWh).[33] The obligatory offshore grid development plan ("*Netzentwicklungsplan Strom Offshore – O-NEP*") must be developed by the German TSOs, and it must be authorized by both, the BNetzA and the BSH.[34] The EEG 2012 introduced new remuneration specifics, and, via the EEG 2014, the offshore expansion targets were modified. The EEG 2014 plans for 6.5 GW instead of 10 GW in 2020 and 15 GW instead of 25 GW in 2030.[35]

Offshore wind energy is legally defined:[36]

> "offshore wind energy installation" shall mean every installation to generate electricity from wind energy which has been erected at sea at a distance of at least three nautical miles measured seawards from the coastline; the coastline shall be taken to be the coastline depicted in Map Number 2920 German North Sea Coast and Adjacent Waters, 1994 edition, XII., and in Map Number 2921 German Baltic Coast and Adjacent Waters, 1994 edition, XII. of the Federal Maritime and Hydrographic Agency, scale of 1: 375,000.[37]

In the following text, the legal framework for offshore wind power in the German EEZ is analyzed, which includes planning, licensing and potential appeal procedures as well as financial support mechanisms for offshore wind power within the framework of the EEG. Section §1.02 addresses the institutional design for offshore wind energy. Section §1.03 focuses on the legal design for offshore wind energy in Germany, including financing schemes, regulation on planning and licensing as well as topics such as SEA, EIA and OWF grid connection, and section §1.04 looks at challenges and solutions regarding offshore wind energy development in Germany.

31. *Pfeil & Töpfer,* Neuregelung für die Genehmigung von Offshore-Windkraftanlagen und Leitungssystemen, NordÖR 9/2011, 373–378; *Büllesfeld, Koch & v. Stackelberg,* Das neue Zulassungsregime für Offshore-Windenergieanlagen in der ausschließlichen Wirtschaftszone (AWZ), ZUR 2012, 274–281; *Spieth & Uibeleisen,* Neues Genehmigungsregime für Offshore-Windparks, NVwZ 2012, 321–325.
32. *BSH,* Standard Investigation of the Impacts of Offshore Wind Turbines on the Marine Environment (StUK4), 2013 (http://www.bsh.de/en/Products/Books/Standard/7003eng.pdf) (accessed November 26, 2016).
33. Sections 17e, 17f EnWG.
34. Sections 17b, c EnWG.
35. Section 3 No. 2 EEG 2014.
36. Section 5 No. 36 EEG 2014.
37. *See* the translation provided by BMWi under http://www.bmwi.de/English/Redaktion/Pdf/renewable-energy-sources-act-eeg-2014,property=pdf,bereich=bmwi2012,sprache=en,rwb=true.pdf (accessed November 26, 2016).

§1.02 INSTITUTIONAL DESIGN FOR OFFSHORE WIND ENERGY

[A] The "Main" Authority for the Development of Offshore Wind Power

The BSH is the relevant authority for the entire process of licensing and operating OWFs in the EEZ.[38] The law for permitting offshore wind-energy installations differs, depending on the wind farms' locations. The territorial sea belongs to the coastal states' sovereign territory. Hence, wind farms within the German 12-Miles-Zone are subject to national law.[39]

In Germany, the conditions for OWFs are different compared to the situation in other successful offshore wind-energy countries such as Denmark or the UK. German OWFs are mainly located in the German EEZ.[40] Designated protected areas (e.g., the National Parks German Wadden Sea) and shipping routes alongside the German coastline create massive obstacles for the installation of wind turbines. Even within the German EEZ, offshore wind areas are limited because approximately one-third of the EEZ are marine-reserve and Natura-2000 areas.[41] Offshore wind-farm projects must therefore keep long distances to the shore (usually at least 40 km), and they must consider relatively deep waters and high investment volumes.[42]

In 2012, the SeeAnlV was amended by a new plan approval procedure for OWFs. Before this amendment, OWFs required a permit in the form of a "conditional decision" ("*gebundene Entscheidung*"), meaning that the BSH had no discretion to decide whether to grant a permit or not. Now, the BSH must balance different interests affected by the wind farm project in a weighing-up procedure. One advantage of this procedure is the so-called concentration effect: the approval includes all permits required for the realization of the OWF,[43] while the preconditions remain the same. The BSH ensures that all relevant acts, requirements and processes are carried out. Authorities such as the Federal Environmental Agency ("*Umweltbundesamt – UBA*")[44] or the Federal Agency for Nature Conservation ("*Bundesamt für Naturschutz – BfN*")[45] are informed about the projects and asked to deliver an opinion.[46] Nature conservation

38. Sections 1, 1a SeeAnlV.
39. *See Petersen & Thomas*, Energy Law in Germany (2014), 254–255.
40. *See* United Nations Convention on the Law of the Sea; Proklamation über die Errichtung einer AWZ der Bundesrepublik Deutschland in der Nordsee und in der Ostsee, November 25, 1994, Federal Law Gazette II, p. 3370.
41. *Fest*, Die Errichtung von Windenergieanlagen in Deutschland und seiner Ausschließlichen Wirtschaftszone (2010), pp. 418–419.
42. *Reichardt* et al., Analyzing interdependencies between policy mixes and technological innovation systems: The case of offshore wind in Germany, in: Technological Forecasting & Social Change 106 (2016), 11–21 (14); *Prall*, in Altrock, Oschmann & Theobald, EEG, (2013), section 31, paras 8–10.
43. *Petersen & Thomas*, Energy Law in Germany (2014), 255.
44. *See* https://www.umweltbundesamt.de/en (accessed November 26, 2016).
45. *See* https://www.bfn.de/index + M52087573ab0.html (accessed November 26, 2016).
46. Section 73 para. 2 VwVfG (Administrative Procedure Act – Verwaltungsverfahrensgesetz, VwVfG of January 23, 2003, Federal Law Gazette I p. 102; *see* the English version under http://www.bmi.bund.de/SharedDocs/Downloads/EN/Gesetzestexte/VwVfg_en.pdf?__blob=publicationFile (accessed November 26, 2016).

agencies participate in the plan approval procedure,[47] and the same applies to environmental protection associations.[48]

The Federal Ministry for Economic Affairs and Energy ("*Bundesministerium für Wirtschaft und Energie – BMWi*")[49] is responsible for the development of the government's energy policy. It oversees the development of renewable energy sources, for instance by preparing alterations of support schemes such as the EEG or by issuing energy-related ordinances. The feasibility of offshore wind power projects still depends on the financial promotion provided by the EEG.

[B] "Supplementary" Authorities for the Development of Offshore Wind Power

[1] Federal Level

The plan approval process includes several authorities. Although the BSH governs the approval process and is the official hearing, planning and licensing authority,[50] various other authorities participate in this process. Authorities whose spheres of competence are affected by the offshore wind energy project shall report their opinions with regard to the project.[51] Regularly, the Federal Environmental Agency, the Federal Agency for Nature Conservation, the military administration and the General Waterways and Shipping Directorates ("*Wasserstraßen- und Schifffahrtsverwaltung des Bundes – WSV*")[52] are affected by OWFs.[53] Nature conservation agencies[54] and environmental protection associations[55] must also be integrated in the plan approval procedure.

Furthermore, the Federal Network Agency is an important authority for offshore wind energy projects. Its competences are mainly stipulated in the Energy Industry Act ("*EnWG*"), concerning *inter alia* access to power supply grids. This includes tasks such as the approval of the TSO's scenario-framework drafts[56] or the development of the Federal Offshore Wind Park Plan (OWP, "*Bundesfachplan Offshore*").[57] The TSOs present the offshore network development plan to the BNetzA for its approval.[58] The

47. Section 63 para. 1 No. 3 BNatSchG (Act on Nature Conservation and Landscape Management – Federal Nature Conservation Act – BNatSchG of July 29, 2009, Federal Law Gazette 2009, I, No. 51, p. 2542 ff.); *see* the English version under http://www.bmub.bund.de/fileadmin/Daten_BMU/Download_PDF/Naturschutz/bnatschg_en_bf.pdf (accessed November 26, 2016).
48. Section 2 para. 6 in conjunction with section 9 para. 1 UVPG (Environmental Impacts Assessment Act -Gesetz über die Umweltverträglichkeitsprüfung – UVPG of February 24, 2010, Federal Law Gazette I p. 94) and section 73 paras 3–7 VwVfG.
49. *See* http://www.bmwi.de/EN/root.html (accessed November 26, 2016).
50. *See* section 2 para. 2 SeeAnlV.
51. Sections 73 para. 2, 73 para. 3a S. 2 VwVfG.
52. Section 8 SeeAnlV; *see also* https://www.wsv.de/ (accessed November 26, 2016).
53. *Hinsch*, in: Schulz, Handbuch Windenergie (2015), 404.
54. Section 63 para. 1 No. 3 BNatSchG.
55. Section 2 para. 6 in conjunction with section 9 para. 1 UVPG, section 73 paras 3–7 VwVfG.
56. Section 12 para. 3 EnWG.
57. In cooperation with BSH, section 17a para. 1 EnWG.
58. Section 17b EnWG.

competencies of the Federal Network Agency are expanded by the "Offshore Wind Sea Act" (*Windenergie-auf-See-Gesetz*) – now, 2016, the Agency is also the relevant authority for the tendering process for new designated offshore areas.

[2] The Role of Regional Governments

Wind farms and the associated offshore cables within the German 12-Miles-Zone are subject to national law. The territorial sea belongs to the coastal state's sovereign territory. In the German federal system, the power is distributed between the federal state ("*Bund*") and the sixteen provinces ("*Länder*"). The Länder Lower Saxony, Schleswig-Holstein and Mecklenburg-Western Pomerania established Regional Development Plans ("*Landesentwicklungspläne, Landes-Raumordnungsprogramm – LROP*"), including the spatial development of the territorial sea for wind-energy use.[59] The Lower Saxony-LROP, for example, aims to promote the use of renewable energy, especially offshore wind energy and the harbors (Cuxhaven and Emden). To avoid negative impacts on tourism and the landscape, the LROP limits the use of offshore wind energy in the territorial sea to a distance of 10 km between offshore wind turbines and the shore or islands. Some offshore areas such as Riffgat and Nordergründe have been designated for wind-energy testing purposes. The designations of suitable offshore areas exclude other areas for offshore wind-energy use within the 12-Miles-Zone.[60]

As part of the German energy transition, offshore wind energy is strongly supported by Länder-governments. For instance, the 2012 Lower Saxony energy concept addresses the necessity and expansion of grid connection for offshore wind farms (especially in the EEZ) and of testing wind-farm sites in the territorial sea. Lower Saxony also supports the development of the offshore harbors Cuxhaven and Emden and addresses the need for qualified, properly trained personnel in the offshore business. It also supports the *ForWind* study program "Offshore Windenergie"[61] and other research facilities.[62]

---

59. Landesentwicklungsplan Schleswig-Holstein 2010, https://www.schleswig-holstein.de/DE/Fachinhalte/L/landesplanung_raumordnung/Downloads/landesentwicklungsplan/landesentwicklungsplan_sh_2010.pdf?__blob=publicationFile&v=5 (accessed November 26, 2016); Landes-Raumordnungsprogramm Niedersachsen 2012, http://www.ml.niedersachsen.de/themen/raumordnung_landesentwicklung/landesraumordnungsprogramm/landes-raumordnungsprogramm-niedersachsen-5062.html (accessed November 26, 2016).
60. Landes-Raumordnungsprogramm Niedersachsen (2012), Anlage 1, http://www.ml.niedersachsen.de/themen/raumordnung_landesentwicklung/landesraumordnungsprogramm/landes-raumordnungsprogramm-niedersachsen-5062.html (accessed November 26, 2016).
61. *See* http://www.offshore-wind-studies.com/index.php/de/?article_id=6&clang=0 (accessed November 26, 2016).
62. *Niedersächsisches Ministerium für Umwelt, Energie und Klimaschutz* (2012), Das Energiekonzept des Landes Niedersachsen, 19–21, 63–64, http://www.netzausbau-niedersachsen.de/downloads/20120131-das-energiekonzept-des-landes-nieders.pdf (accessed November 26, 2016).

[3] Other Players Vital for the Development of Offshore Wind

Apart from state-based regulation and promotion, various private players contribute to the development of offshore wind-energy use in Germany. In the first years of the new century, German manufacturers developed advanced types of offshore wind turbines.[63] In 2005, the "Foundation Offshore Wind Energy" was established to promote wind energy at sea.[64] Mainly financed by the BMU, it represents various stakeholder interests and seeks to improve research on offshore wind energy use. In 2010, "Alpha Ventus"[65] was commissioned as the first German OWF, serving as a test field for the foundation. The BMU provided EUR 30 million of funds for "Alpha Ventus," with a total investment volume of EUR 250 million.[66] The BMU (now the BMWi) promoted various research projects within the framework of the so-called RAVE research initiative. The research projects focused on the test site "Alpha Ventus," generating experience and knowledge for new wind farms to come. The BMWi provided EUR 50 million for the RAVE-research programs.[67]

[C] The Role of "Law" Within Institutional Decision Making

The role of law can be seen from various sides. On the one hand, law can retrospectively react to real world developments by trying to govern unwanted effects. German and European energy law has reacted in this respect to technological and economic developments in the field of renewable energies. One example was the falling prices for solar panels due to technological innovation and cheap imports from the People's Republic of China, mainly from 2010 to 2012. German legislature reacted hastily by reducing feed-in-tariffs provided by the EEG. This was, however, too late - new installations in these years climbed to 7,500 MW/a. With respect to the twenty-year-guarantee for the tariffs, this led to solar debts of several billion Euros which must be paid by the consumer via the so-called EEG-surcharge - in a country with only suboptimal conditions for solar energy.[68] Regarding OWFs, a reverse development can be found. Here, prices for the installation of offshore wind turbines rose, leading to the

63. *Reichardt* et al., Analyzing interdependencies between policy mixes and technological innovation systems: The case of offshore wind in Germany, Technological Forecasting & Social Change 106 (2016), 11 (15).
64. *Reichardt* et al., Analyzing interdependencies between policy mixes and technological innovation systems: The case of offshore wind in Germany, Technological Forecasting & Social Change 106 (2016), 11 (16).
65. *See* https://www.alpha-ventus.de/english/ (accessed November 26, 2016).
66. *See* https://www.alpha-ventus.de/fileadmin/Dateien/publikationen/av_Factsheet_de_2016.pdf (accessed November 26, 2016).
67. *See* Research at Alpha Ventus (RAVE), 2016 (http://rave.iwes.fraunhofer.de/rave/pages/welcome) (accessed November 26, 2016).
68. *Schomerus*, Renewable energy: support mechanisms, in Farber &Peeters (editors), Climate Change Law Vol. 1, S. 465, 468, Elgar Encyclopedia of Environmental Law, edited by Michael Faure, 2016.

raising of feed-in-tariffs as a legislative reaction. For example, the 2014 amendment of the EEG saw the basic tariff increased from 3.5 to 3.9 Cent/kWh.[69]

On the other hand, law can be prospectively used, as an instrument for changing future society.[70] The best example in the field of renewable energies is the total regime change triggered by the 2014 EU Commission's guidelines.[71] Hitherto, EU Member States could more or less freely decide which system to choose for promoting renewable energies, feed-in tariffs, feed-in premiums, quota obligations or tendering and auction schemes.[72] From 2021 onwards, aid may only be "granted in a competitive bidding process on the basis of clear, transparent and nondiscriminatory criteria."[73] By changing the law, operators are now forced to compete with each other, leading to a more market-based promotion scheme. To implement this obligation, German legislature passed a new law, the so-called WindSeeGesetz.[74] One of its main purposes is to introduce a tender and auctioning scheme for the installation of OWFs.[75]

One of the reasons for the new promotion scheme was the relative inertia in the former system of guaranteed and legally fixed feed-in-tariffs. As the solar-panel example shows, the German parliament's reaction was not quick enough to meet the needs of the renewable energy market. The same applies to the offshore sector: a competitive bidding scheme can react much quicker to technological and economic changes than the slow and complex legal and political processes of changing federal law.

A third area in which law can influence actual developments is jurisdiction. Other than in the case of onshore wind farms, only a few court judgments can be found in the offshore sector. Standing against OWFs in the EEZ is limited.[76] The rescission of an administrative act can be requested by means of an action according to the Code of Administrative Court Procedure ("*Verwaltungsgerichtsordnung – VwGO*").[77] The claim is only admissible if the plaintiff claims a violation of his rights by either the

69. *Schomerus &Meister*, section 50, in Frenz/Müggenborg/Cosack/Ekardt (ed.), Erneuerbare Energien Gesetz, 2015, p. 1175.
70. *See Peeters & Schomerus*, Modifying Our Society With Law: The Case of EU Renewable Energy Law, Climate Law 4 (2014) 131.
71. *EU Commission*, Guidelines on State aid for environmental protection and energy 2014–2020 (2014/C 200/01), available under http://eur-lex.europa.eu/legal-content/EN/TXT/PDF/?uri=CELEX:52014XC0628(01)&from=EN (accessed November 26, 2016).
72. *See* the overview at *Schomerus*, Renewable energy: support mechanisms, in Farber &Peeters (editors), Climate Change Law Vol. 1, S. 465 et seq., Elgar Encyclopedia of Environmental Law, edited by Michael Faure, 2016.
73. *EU Commission*, Guidelines on State aid for environmental protection and energy 2014–2020 (2014/C 200/01), p. 26, available under http://eur-lex.europa.eu/legal-content/EN/TXT/PDF/?uri=CELEX:52014XC0628(01)&from=EN (accessed November 26, 2016).
74. Gesetz zur Entwicklung und Förderung der Windenergie auf See of October 13, 2016, Federal Law Gazette I p. 2258.
75. *See* sections 15 et seq. WindSeeG.
76. *Gatz*, Windenergieanlagen in der Verwaltungs- und Gerichtspraxis (2013), 242; *Fest*, Die Errichtung von Windenergieanlagen in Deutschland und seiner Ausschließlichen Wirtschaftszone (2010), 462.
77. Code of Administrative Court Procedure of March 19, 1991, Federal Law Gazette I p. 686; English version available under https://www.gesetze-im-internet.de/englisch_vwgo/englisch_vwgo.html (accessed November 26, 2016).

administrative act or its refusal or omission.[78] Claims of offshore project developers are usually admissible, as they are normally affected by a (negative) decision of the BSH.[79]

For private third parties, however, it is difficult to prove a violation in their subjective rights, since no "local residents" exist in the vicinity of OWFs. The OWFs in the German North Sea-EEZ are generally located far away from the German coast or islands. Local residents are usually not affected by adverse effects of an OWF. Regarding claims against OWFs in the EEZ, German jurisdiction has been very strict and dismissive. Private third party claims against OWP are usually not admissible.[80] For instance, the German Supreme Court denied a violation of rights regarding a professional deep-sea fishing company by an OWF.[81] The purpose of section 5 paragraph 6 SeeAnlV is to protect public rather than individual interests. An infringement of a fundamental right as stipulated in Article 12 or in Article 14 German Basic Law ("*Grundgesetz – GG*")[82] is very unlikely, as the plan approval of an OWF will not restrict the fishermen's profession. The actual limitation of fishing grounds due to the location of the wind farm is marginal compared to the whole EEZ. Furthermore, the potential ability to fish in a specific area is not covered by Article 14 GG.

In the same way, municipality rights are unlikely to be infringed by OWFs.[83] Due to the location of the OWFs in the EEZ, negative effects on tourism are not to be expected. Further, island municipalities' possible defense rights do not cover indirect effects such as potential negative effects on tourism.[84]

NGOs aimed at nature conservation are entitled to file claims against OWFs on the grounds of nature protection.[85] Moreover, NGOs aiming at environmental protection in general may file claims against decisions in the context of the projects' Environmental Impact Assessment (EIA);[86] a violation in their subjective rights is not required.[87]

The situation regarding connection cables for OWFs in the territorial sea is different. Such cables are subject to plan approval procedures, which enables third parties to file claims against them.[88] In such cases, individuals must proof a violation of their subjective rights to be held admissible.[89]

78. Section 42 para. 2 VwGO.
79. *Hinsch*, in: Schulz, Handbuch Windenergie (2015), 418–419, with further references.
80. *Hinsch*, in: Schulz, Handbuch Windenergie (2015), 419.
81. *BVerfG*, Non-acceptance decision of April 26, 2010 – 2 BvR 2179/04, NVwZ 2010, 555; OVG Hamburg, Decision of September 30, 2004 – 1 Bf 162/04 – juris.
82. Basic Law for the Federal Republic of Germany, as amended by Article 1 of the Act of December 23, 2014, Federal Law Gazette I p. 2438; English version available under https://www.gesetze-im-internet.de/englisch_gg/ (accessed November 26, 2016).
83. *BVerfG*, Non-acceptance decision of November 12, 2009 – 2 BvR 2034/04 -juris; OVG Hamburg, decision of September 15, 2004 – 1 Bf 128/04, NVwZ 2005, 347.
84. *OVG Lüneburg*, decision of September 13, 2010 – 12 LA 18/09 – ZNER 2010, 511 (513).
85. Sections 64 para. 1, 63 para. 1 No. 3 BNatSchG; *see Gatz*, Windenergieanlagen in der Verwaltungs- und Gerichtspraxis (2013), 249–251.
86. Section 2 para.1 UmwRG (Gesetz über ergänzende Vorschriften zu Rechtsbehelfen in Umweltangelegenheiten nach der EG-Richtlinie 2003/35/EG – Umwelt-Rechtsbehelfsgesetz – UmwRG of April 8, 2013, Federal Law Gazette I p. 753).
87. *Hinsch*, in: Schulz, Handbuch Windenergie (2015), 421.
88. *See* section 43 sentence 1 No. 3 EnWG.
89. *Petersen & Thomas*, Energy Law in Germany (2014), 257.

§1.03 LEGAL DESIGN FOR OFFSHORE WIND

[A] Incentives

[1] Main Finance Scheme

Support mechanisms for the offshore-wind-energy sector are indispensable if the ambitious expansion targets are to be reached. There is a broad range of such mechanisms reaching from feed-in-tariffs over feed-in-premiums, quota obligations such as renewable portfolio standards, and tendering and auction schemes, to fiscal incentives.[90] As mentioned above, regarding the German support system, the year 2020 marks a shift from a feed-in-tariff, respectively a feed-in-premium, to a tendering and auction scheme.

Up to 2020, the EEG 2014 offers two models for OWFs, both on the basis of the feed-in-premium scheme. For both models, the support period is twenty years:

> The financial support must in each case be paid for the duration of 20 calendar years plus the year of commissioning of the installation. The commencement of the period pursuant to sentence 1 shall be the date of the commissioning of the installation unless the following provisions state otherwise.[91]

According to the basic model, section 50 EEG guarantees a value of 15.40 cents per kilowatt-hour for the first twelve years (so-called initial value). The initial value period can be extended, depending on the OWFs' location and the water depth:

> The period pursuant to sentence 1 shall be extended by 0.5 months for each full nautical mile extending beyond twelve nautical miles which the installation is further away from the coastline pursuant to Section 5 number 36 second half-sentence, and by 1.7 months for every full metre of water depth exceeding a depth of 20 metres. The water depth shall be determined on the basis of the chart datum.[92]

Once the conditions for this so-called initial value cease to exist, the value to be applied goes down to 3.90 cents/kWh, the so-called basic value.

A second way of support offered by the EEG is the so-called acceleration model. The wind farm (if commissioned before January 1, 2020) will receive an initial value of 19.4 cents/kWh for the first eight years from the commissioning of the installation:

---

90. *See Schomerus*, Renewable energy: support mechanisms, in Farber & Peeters (editors), Climate Change Law Vol. 1, S. 465 et seq., Elgar Encyclopedia of Environmental Law, edited by Michael Faure, 2016.
91. Section 22 EEG 2014; *see* the translation provided by BMWi under http://www.bmwi.de/English/Redaktion/Pdf/renewable-energy-sources-act-eeg-2014,property = pdf,bereich = bmwi2012,sprache = en,rwb = true.pdf (accessed November 26, 2016).
92. Section 50 para. 2 sentences 2 and 3 EEG 2014; *see* the translation provided by the BMWi under http://www.bmwi.de/English/Redaktion/Pdf/renewable-energy-sources-act-eeg-2014,property = pdf,bereich = bmwi2012,sprache = en,rwb = true.pdf (accessed November 26, 2016).

> If the offshore wind energy installation is commissioned before 1 January 2020 or its operational readiness has been established under the preconditions of Section 30 subsection 2, the value to be applied shall in derogation of subsection 1 amount to 19.40 cents per kilowatt-hour in the first eight years from commissioning of the installation if the installation operator demands this from the grid system operator before the commissioning of the installation. In this case, the entitlement pursuant to subsection 2 sentence 1 shall be lost, whilst the entitlement to the payment pursuant to sub-section 2 sentence 2 shall be applied mutatis mutandis with the proviso that the initial value shall amount to 15.40 cents per kilowatt-hour in the period of the extension.[93]

One must, however, consider that the support scheme is accompanied by a system of gradual reduction, meaning that dependent on the commissioning year the tariff will be reduced. There is a specific reduction rate for each technology. Section 30 EEG 2014 marks the rates for OWFs:

> Gradual reduction of the support for electricity from offshore wind energy
> (1) The values to be applied shall be reduced for electricity from offshore wind energy
> 1. pursuant to Section 50 subsection 2
>   a) by 0.5 cents per kilowatt-hour from 1 January 2018,
>   b) by 1.0 cent per kilowatt-hour from 1 January 2020,
>   c) by 0.5 cents per kilowatt-hour each year on 1 January from 2021,
> 2. pursuant to Section 50 subsection 3 by 1.0 cent per kilowatt-hour from 1 January 2018.[94]

For wind farms located in protected nature and landscape areas built after December 31, 2004, financial support is excluded:

> Subsections 1 to 4 shall not be applied to offshore wind energy installations whose erection has been authorised in an area of the German exclusive economic zone or the coastal sea after 31 December 2004 which has been declared a protected part of nature and landscape pursuant to Section 57 in conjunction with Section 32 subsection 2 of the Federal Nature Conservation Act or under Länder legislation. Sentence 1 shall also apply until the protection is imposed for those areas which the Federal Ministry for the Environment, Nature Conservation, Building and Nuclear Safety has nominated to the European Commission as areas of Community importance or as European bird sanctuaries.[95]

From 2021 onwards, an auctioning scheme with a competitive bidding process will replace the former feed-in-premium system. The BMWi gave the reasoning behind the new rules:

93. Section 50 para. 3 EEG 2014; *see* the translation provided by BMWi under http://www.bmwi.de/English/Redaktion/Pdf/renewable-energy-sources-act-eeg-2014,property=pdf,bereich=bmwi2012,sprache=en,rwb=true.pdf (accessed November 26, 2016).
94. *See* the translation provided by BMWi under http://www.bmwi.de/English/Redaktion/Pdf/renewable-energy-sources-act-eeg-2014,property=pdf,bereich=bmwi2012,sprache=en,rwb=true.pdf (accessed November 26, 2016).
95. Section 50 para. 5 EEG 2014, *see* the translation provided by BMWi under http://www.bmwi.de/English/Redaktion/Pdf/renewable-energy-sources-act-eeg-2014,property=pdf,bereich=bmwi2012,sprache=en,rwb=true.pdf (accessed November 26, 2016).

> rather than being fixed by the government, future rates of renewables funding will be determined by the market by means of dedicated auction schemes. This is because renewables have matured and are now mature enough to compete on the market. The new auction scheme is to ensure that the expansion of renewables proceeds at a steady and controlled pace and at a low cost. The legislation also enables us to make sure that the high level of market-player diversity that has characterised the energy transition will be upheld.[96]

The new rules for financial support follow the so-called central "Danish" target model. According to the BMWi, its cornerstones are as follows:

- Government examines in advance the sites to be auctioned for wind farms. This ensures optimal dovetailing with the grid connections.
- In every other model, a stock of grid connections would have to be built. Otherwise there would be no competition. This would entail massive extra costs.
- Until the new model is introduced in 2026, auctions will take place on a transitional basis amongst the wind farms already in planning. It will be ensured that there is no sudden interruption to the development of the industry after 2020.[97]

Primarily, the WindSeeG is the relevant act to be applied, but alternatively the EEG 2017 with its more general provisions is also applicable. The WindSeeG provides in its sections 14 et seq. the conditions for granting the market premium under a competitive procedure. Section 14 says that OWF-operators of installations commissioned after December 31, 2020 may only claim financial support by the feed-in-premium scheme once their bid is accepted. From 2021, on September 1 of each year, a tender volume of 700–900 MW will be advertised. The tender is based on an Offshore Development Plan ("*Flächenentwicklungsplan*") made by the BSH in cooperation with the BNetzA. The plan is designed to specify areas for OWFs, the chronological order of the tender procedures for the selected areas, the years of the installation of the respective OWF, the cable connections, etc. It must also consider regional planning requirements, maritime environmental impacts, maritime traffic and military requirements. Potential bidders must be given sufficient information concerning the area for which the specific tender will be advertised. Bidders must deposit a security, and a maximum value for the bids must be determined.

The provisions for the auctioning under the EEG 2017 and the WindSeeG are full of complex details. It is not yet quite clear whether the auctioning scheme will succeed in practice. First experience with the so-called Pilot-Auctioning-Scheme for ground-mounted solar installations was relatively positive. However, conditions for OWFs are much different from those for solar panels.

---

96. *BMWi*, 2017 Renewable Energy Sources Act adopted: Ringing in the next phase of the energy transition, 2016 (http://www.bmwi.de/EN/Topics/Energy/Renewable-Energy/renewable-energy-sources-act-2017.html) (accessed November 26, 2016).
97. *BMWi*, 2017 revision of the Renewable Energy Sources Act, 2016 (http://www.bmwi.de/English/Redaktion/Pdf/eeg-novelle-2017-praesentation,property=pdf,bereich=bmwi2012,sprache=en,rwb=true.pdf) (accessed November 26, 2016).

[2] Supplementary Finance Scheme

[a] Low Interest Loans and Loan Guarantees

Offshore wind-energy projects are characterized by significant investment volumes.[98] This applies all the more to the German situation where OWF-locations in the EEZ lie far away from the shore. Governmental finance schemes and incentives are needed. An important instrument is the loan program "Offshore Wind Energy" provided by the state-owned KfW-bank of 2011,[99] offering investment credits for OWFs up to EUR 500 million per project. The program grants loans at market conditions (total volume of EUR 5 billion) for up to ten German OWFs. It was introduced against the background of the financial crisis and grid-connection problems,[100] and it is intended to close financing gaps for offshore projects in the banking sector.[101] Project companies investing in the EEZ or in the 12-Mile-Zone of the North Sea or the Baltic Sea are eligible to apply for the project financing. As a condition for being financed under the "Offshore Wind Energy Scheme," wind farms must meet the EEG 2014 requirements.[102] As of August 2016, five offshore wind farms have used the KfW program: "Meerwind (KfW-volume €264 million),"[103] "Global Tech I (KfW-volume around € 280 million),"[104] "Butendiek (KfW- and KfW IPEX-Bank-volume €239 million),"[105] "Veja Mate (KfW-volume around €430 million)"[106] and "Merkur (KfW-volume over €360

98. *Winkelmann*, Die Welt der großen Zahlen, in: Bundesverband WindEnergie e. V., Finanzierung Onshore & Offshore (2012), 86–87.
99. *Kreditanstalt für Wiederaufbau; see Winkelmann*, Die Welt der großen Zahlen, in: Bundesverband WindEnergie e. V., Finanzierung Onshore & Offshore (2012), 86–87.
100. http://www.offshore-stiftung.com/60005/Uploaded/BvBuelow_R_090611.pdf (accessed June 28, 2016); Bundesregierung (2010), Energiekonzept, 9, accessible at: https://www.bundesregierung.de/ContentArchiv/DE/Archiv17/_Anlagen/2012/02/energiekonzept-final.pdf?__blob=publicationFile&v=5 (accessed November 26, 2016).
101. *Winkelmann*, Die Welt der großen Zahlen, in: Bundesverband WindEnergie e. V., Finanzierung Onshore & Offshore (2012), 87.
102. *Kreditanstalt für Wiederaufbau*, Offshore Wind Energy Programme, 2016 (https://www.kfw.de/Download-Center/F%C3%B6rderprogramme-(Inlandsf%C3%B6rderung)/PDF-Dokumente/6000002171-M-Offshore-Windenergie-englisch.pdf) (accessed November 26, 2016).
103. *See Kreditanstalt für Wiederaufbau* (2016) https://www.kfw.de/KfW-Konzern/Newsroom/Aktuelles/Pressemitteilungen/Pressemitteilungen-Details_10227.html# (accessed November 26, 2016).
104. *See Kreditanstalt für Wiederaufbau* (2016) https://www.kfw.de/KfW-Konzern/Newsroom/Aktuelles/Pressemitteilungen/Pressemitteilungen-Details_10211.html# (accessed November 26, 2016).
105. *See Kreditanstalt für Wiederaufbau* (2016) https://www.kfw.de/inlandsfoerderung/Technische-Seiten/Pressmitteilung-Details_20006-2.html (accessed November 26, 2016).
106. *See Kreditanstalt für Wiederaufbau* (2016) https://www.kfw.de/KfW-Group/Newsroom/Aktuelles/Pressemitteilungen/Pressemitteilungen-Details_285888.html (accessed November 26, 2016).

million)."[107] Furthermore, the KfW-IMPX Bank offers special loan guarantee arrangements for offshore wind farms.[108]

[b] Tax Credits

Profits of renewable energy projects are basically subject to corporate tax / income tax and business tax. Since 2008, this also applies to OWFs in the EEZ. There are no special corporate or income tax regulations concerning OWFs.

The situation is different with regard to business tax ("*Gewerbesteuer*"). Generally, this tax is levied by the municipalities, e.g., in the case of onshore wind-energy projects. However, the EEZ does not belong to any municipality. Here, according to section 4(2) German Trade Act ("*Gewerbesteuergesetz – GewStG*"),[109] the state governments possess the right to levy the tax. For instance, the state of Lower Saxony holds the right of taxation (business tax of 14.7% in 2014). The state of Schleswig-Holstein has transferred the right of taxation to the (island-)municipality of Heligoland (business tax of 12.25% in 2014), and the state of Mecklenburg-Western Pomerania has set up a business tax of 13.72% for the EEZ in the Baltic Sea in 2014.[110]

[c] Favorable Grid Cost-Sharing Rules

Providing sufficient grid connection for OWFs is still a big challenge.[111] The TSOs (*TenneT TSO GmbH* for the North Sea and *50Hertz Transmission GmbH* for the Baltic Sea) must establish the grid connection according to the Offshore Network Development Plan (O-NEP).[112] The TSOs must develop the O-NEP on the basis of the scenario-framework (also developed by TSOs) and the "*Bundesfachplan Offshore*" (developed by BSH). The O-NEP covers the development of grid connection for OWFs within the next ten years.[113] The plan must be approved by the BNetzA, in coordination with the BSH.[114] The O-NEP follows a cluster-connection concept, covering the realization-date for the grid connection, the location and its size.[115] The plan becomes

107. *See Kreditanstalt für Wiederaufbau* (2016) https://www.kfw.de/KfW-Group/Newsroom/Aktuelles/Pressemitteilungen/Pressemitteilungen-Details_369728.html (accessed November 26, 2016).
108. For instance the OWFs Veja Mate and Merkur, *see*: https://www.kfw-ipex-bank.de/International-financing/KfW-IPEX-Bank/Presse/News/Newsdetails_286656.html, https://www.kfw.de/nachhaltigkeit/News/News-Details_369600.html (accessed November 26, 2016).
109. Gewerbesteuergesetz of October 15, 2002, Federal Law Gazette I p. 4167.
110. *Ropohl*, Besteuerung von Erneuerbare-Energien-Projekten in Deutschland, in: Gerhard, Rüschen & Sandhövel, Finanzierung Erneuerbarer Energien (2015), 145.
111. *Schulz*, Netzanschluss, in: Schulz, Handbuch Windenergie (2015), 77–78.
112. Section 17d para. 1 EnWG.
113. BMWi, Offshore-Windenergie (2015), p. 7 (https://www.bmwi.de/BMWi/Redaktion/PDF/Publikationen/offshore-windenergie,property = pdf,bereich = bmwi2012,sprache = de,rwb = true.pdf) (accessed November 26, 2016).
114. Section 17c EnWG.
115. Section 17b EnWG.

binding for the TSOs thirty months before the expected grid connection completion.[116] The BNetzA allocates the grid capacity from each cluster to the wind farms. If a wind farm completion is delayed, a *use- it-or-lose-it-principle* is applied. In such cases, the BNetzA may reallocate the capacity to another wind farm within the cluster.[117]

The TSOs can shift the grid-connection costs to the network charges.[118] In the past, liability and compensation issues were not clearly regulated, resulting in OWF grid expansion delays.[119] In 2012, a new liability provision was adopted. If, due to a missing or defective grid connection, an OWF cannot feed electricity into the grid, from the 11th day of the delay on, TSOs are liable for liquidated damages. The project developers are entitled to 90% of the potential EEG remuneration.[120] TSOs can shift the costs (up to 0.25 cent/kWh) of liquidated damages to the final electricity consumer (the so-called offshore-liability surcharge – "*Offshore-Haftungsumlage*").[121]

[d] Links to ETS

The Emissions Trading Scheme (ETS) was introduced in Germany in 2005. However, in the first emissions trading period from 2005–2007, it failed due to an over-allocation of certificates. The outcome was additional revenues for the operators of coal-fired power stations and higher electricity prices for consumers, but next to no CO_2 reductions.[122] Considering the current price of EUR 5.27/ton CO_2 for EU Emission Allowances,[123] it is still far too low. Although the Commission considers the increase in renewables (including offshore wind power) when calibrating the ETS cap-reductions, there is still criticism from stakeholders such as the European Wind Energy Association (EWEA):

> Weaker demand for ETS allowances, in particular in the power sector, will result in additional surpluses in the future as the level of ambition in the ETS does not reflect the emissions reductions realised in reality. For many years, the level of emissions reductions in the EU has been outpacing the annual emission allowances decline rate. Emissions in the ETS sectors decreased by 4.5% in 2014, 3% in 2013 and 2% in 2012. This is considerably more than the current linear reduction factor of 1.74%. To align the ETS with the pace of emissions reductions realised in the current trading period and to reinforce market balance and the effectiveness of

116. Section 17d para. 2 S. 5 EnWG.
117. Section 17d para. 6 S. 3, 4 EnWG; *see also Schulz*, Netzanschluss, in: Schulz, Handbuch Windenergie (2015), 104–105.
118. Section 17d para. 7 EnWG; section 28(2), (3) KWGK.
119. *Schulz*, Netzanschluss, in: Schulz, Handbuch Windenergie (2015), 77.
120. Section 17e para. 1 EnWG.
121. Section 17f paras 5 and 6 EnWG.
122. *Schomerus*, German climate and energy legislation: An ambitious but fragmented framework, in Peeters, Stallworthy & de Cendra de Larragan (ed.), Climate Law in EU Member States: Towards national regulatory frameworks for climate change mitigation, Edward Elgar Publishing Limited, 2012, pp. 178, 195.
123. Of November 24, 2016, *see EEX*, https://www.eex.com/de/marktdaten/umweltprodukte/spotmarkt/european-emission-allowances#!/2016/11/24 (accessed November 26, 2016).

the system, the overhang of allowances needs to be curtailed through different means.[124]

The Commission reacted by planning to increase the pace of emissions cuts. From 2021 onwards, the overall number of emission allowances is supposed to decline at an annual rate of 2.2%.[125]

[e] *Other Funding Sources*

A variety of research institutes and research projects are funded by public means. Funding mainly comes from federal sources such as the Federal Ministry of Education and Research (BMBF)[126] or the BMWi, but also from the Länder-governments. Major research centers for wind energy[127] are Fraunhofer IWES (Fraunhofer Institute for Wind Energy and Energy System Technology),[128] ForWind (Centre for Wind Energy Research),[129] EEK.SH (CE WindEnergy SH)[130] and DEWI (Deutsches Windenergieinstitut), with a special focus on offshore wind energy.[131]

[f] *International and European Trade Law Concerns*

The EU Commission considers the EEG-promotion scheme for renewables a state aid. According to Article 107 TFEU,[132] such subsidies are, in principle, incompatible with the internal market. The Commission can, however, allow certain subsidies under the requirements of Articles 107 et seq. TFEU. The Commission's guidelines of 2015 marked the conditions under which the promotion of renewables shall be allowed.[133] One of these conditions is the introduction of a competitive bidding scheme.[134]

---

124. *EWEA*, Position paper of the European Wind Energy Industry on the ETS post-2020 revision, 2015, p. 4 (http://www.ewea.org/fileadmin/files/library/publications/position-papers/Effective-Transition-Signals-EWEA-Position-Paper.pdf) (accessed November 26, 2016).
125. *EU Commission*, Revision for phase 4 (2021–2030), 2016 (https://ec.europa.eu/clima/policies/ets/revision/index_en.htm) (accessed November 26, 2016); *see also* the Commission's proposal for a directive amending Directive 2003/87/EC to enhance cost-effective emission reductions and low carbon investments, COM(2015) 337 final.
126. *See* https://www.bmbf.de/en/index.html (accessed November 26, 2016).
127. *See* the overview at *BMWi*, Forschungszentren, 2016 (http://www.erneuerbare-energien.de/EE/Navigation/DE/Forschung/Windenergie-auf-See/Forschungszentren/forschungszentren.html) (accessed November 26, 2016).
128. *See* http://www.iwes.fraunhofer.de/en.html (accessed November 26, 2016).
129. *See* http://www.forwind.de/forwind/index.php?article_id = 1&clang = 1 (accessed November 26, 2016).
130. *See* http://www.windenergy-sh.de/home.html (accessed November 26, 2016).
131. *See* http://www.dewi.de/dewi_res/index.php?id = 1&L = 0 (accessed November 26, 2016).
132. Treaty on the Functioning of the European Union, Official Journal of the European Union C 326/50 of October 26, 2012, http://eur-lex.europa.eu/legal-content/EN/TXT/PDF/?uri = CELEX:12012E/TXT&from = en (accessed November 26, 2016).
133. *EU Commission*, Guidelines on State aid for environmental protection and energy 2014–2020 (2014/C 200/01), http://eur-lex.europa.eu/legal-content/EN/TXT/PDF/?uri = CELEX:52014XC0628(01)&from = EN (accessed November 26, 2016).
134. *See* above under §1.03[A][1].

Similar problems can arise with regard to WTO-law. In the Ontario-case, the Canadian province had passed a so-called local-content rule, according to which a feed-in-tariff would only be granted to producer of electricity from renewable energy if a certain percentage of the turbine was produced in Ontario. The WTO-Panel[135] as well as the Appellate Body[136] considered this requirement as incompatible with Article III:4 GATT.[137] Apart from the EU Commission, the Appellate Body did not, however, require a certain method for promoting renewables, nor did it require a certain objective followed by the subsidy. This means that WTO-law gives states more leeway in subsidizing renewables than the EU.[138]

[3] Incentives for Construction Harbors and Construction Vessels and Grids, or Turbine Manufacturers

To ensure a successful offshore wind-energy development, an appropriate infrastructure is essential. OWFs depend on adequate harbors for the offshore wind-energy industry. The planning, construction and maintenance of an OWF requires various specialized vessels and platforms (e.g., exploration ships, dredging vessels, hub platforms, jack-up crane vessels, transport vessels, maintenance and reparation vessels). An active offshore-fleet is necessary to construct, maintain and repair offshore wind turbines in time. For reaching the offshore wind-energy targets, harbors and shipyards in the North and Baltic Sea play a key role.[139] In Germany, there are three offshore port bases (Bremerhaven, Cuxhaven and Sassnitz), seven component harbors and three offshore service ports.[140] For instance, since 2007, the state of Lower Saxony has invested over EUR 200 million[141] in the offshore harbor infrastructure of Cuxhaven. Consequently, various offshore-related companies have settled at this location.[142]

135. Canada – Certain Measures Affecting the Renewable Energy Generation Sector, Panel Report of 19 December 2012, WT/DS412/R (complainant Japan); Canada – Measures Relating to the Feed-in Tariff Program, Panel Report of December 19, 2012, WT/DS426/R (complainant EU).
136. Canada – Certain Measures Affecting the Renewable Energy Generation Sector, Appellate Body Report of May 6, 2013, WT/DS412/AB/R; Canada – Measures Relating to the Feed-in Tariff Program, Appellate Body Report of May 6, 2013, WT/DS426/AB/R.
137. General Agreement on Tariffs and Trade 1994 of April 15, 1994, OJEU 1994 L 336 p. 11.
138. *Kahl*, Local-content requirements in renewable energy support schemes from a trade law perspective, in: Squintani/Vedder/Reese/Vanheusden, Sustainable Energy United in Diversity, EELF Book Series Vol. 1 (2014), p. 219.
139. *See BMWi*, http://www.erneuerbare-energien.de/EE/Navigation/DE/Technologien/Windenergie-auf-See/Wirtschaftliche_Aspekte/Haefen/haefen.html; *BMWi*, http://www.erneuerbare-energien.de/EE/Navigation/DE/Technologien/Windenergie-auf-See/Wirtschaftliche_Aspekte/Werften-Schiffbau/werften-schiffbau.html; *BMWi*, http://www.erneuerbare-energien.de/EE/Navigation/DE/Technologien/Windenergie-auf-See/Wirtschaftliche_Aspekte/Schiffstypen/schiffstypen.html (all accessed November 26, 2016).
140. *See Foundation Offshore Wind Energy*, http://www.offshore-stiftung.de/sites/offshorelink.de/files/mediaimages/Map_Status%20Quo_Offshore%20Wind%20in%20Germany_30.6.2016.jpg (accessed November 26, 2016).
141. Together with the EU and the German Federal Government.
142. *See Lower Saxony Ministry for the Environment, Energy, and Climate Protection*, http://www.umwelt.niedersachsen.de/erneuerbare_energien/wind/offshorehaefen/offshore-haefen-121300.html (accessed November 26, 2016).

The "Foundation Offshore"-project "Vernetzung der Martimen Wirtschaft mit der Offshore-Windenergie 2" targets to improve the cooperation between companies working in maritime economy and the offshore wind-energy framework.[143] In the offshore business, various initiatives focus on different sectors and challenges.[144]

[B] Regulations

[1] License Scheme

German OWFs are mainly planned and located in the EEZ.[145] Only within the twelve nautical mile limit, i.e., in the area of the territorial sea, the German coastal states such as Schleswig-Holstein, Lower-Saxony and Mecklenburg-West Pomerania are responsible for the approval. Due to various restrictions such as nature protection laws or shipping requirements, the territorial area plays only a minor role for OWFs. Hence, the following text only applies to OWFs and their grid connections in the EEZ.[146]

Within the EEZ, the BSH is the authority in charge. The approval for an OWF in the EEZ has no binding effect for further necessary installations on land and in the territorial sea. The international legal basis for all activities in the EEZ is the United Nations Convention on the Law of the Sea of December 10, 1982 (UNCLOS).[147] Specific legal requirements for the installation of OWFs can be found in the Federal Maritime Responsibilities Act ("*Seeaufgabengesetz - SeeAufgG*")[148] and mainly in the Marine Facilities Ordinance ("*Seeanlagenverordnung*" - *SeeAnlV*"). The BSH must approve an OWF if it fulfills certain requirements, such as safety and efficiency of navigation and aspects of marine environment. The regional Waterways and Shipping Directorate must give its consent to the OWF-project.

143. *See Foundation Offshore Wind Energy*, http://www.offshore-stiftung.de/Vernetzung (accessed November 26, 2016).
144. *See Foundation Offshore Wind Energy*, http://www.offshore-stiftung.de/sites/offshorelink.de/files/documents/Initiativen_Stand_201502_final.pdf (accessed November 26, 2016).
145. *See* for the following text *BSH*, wind farms, 2016 (http://www.bsh.de/en/Marine_uses/Industry/Wind_farms/index.jsp) and *BSH*, Plan approval procedure for the construction and operation of installations in the EEZ, 2016 (http://www.bsh.de/en/Marine_uses/Industry/Wind_farms/Approval_Procedure.jsp) (accessed November 26, 2016).
146. Wind turbines in the territorial sea (outside the EEZ) with a total height of at least 50 m are subject to a licensing procedure under the Federal Immissions Control Act ("*Bundes-Immissionsschutzgesetz –BImSchG*" - Gesetz zum Schutz vor schädlichen Umwelteinwirkungen durch Luftverunreinigungen, Geräusche, Erschütterungen und ähnliche Vorgänge of May 17, 2013, Federal Law Gazette I p. 1274); Additional requirements are provided in the Water Resources Act ("*Wasserhaushaltsgesetz – WHG*" - Wasserhaushaltsgesetz of July 31, 2009, (Federal Law Gazette I p. 2585) and statutes on the safety of shipping such as the "*Bundeswasserstraßengesetz – WaStrG*" - Bundeswasserstraßengesetz of May 23, 2007 (Federal Law Gazette I p. 962). These regulations do not fall not under the so-called concentration effect of the BImSchG, making further permits beyond the license according to the BImSchG necessary; *see Gatz*, Windenergieanlagen in der Verwaltungs- und Gerichtspraxis (2013), 206–207.
147. *See United Nations*, http://www.un.org/depts/los/convention_agreements/texts/unclos/unclos_e.pdf (accessed November 26, 2016).
148. Seeaufgabengesetz of June 17, 2016, Federal Law Gazette I p. 1489.

There are two different approval procedures. In a licensing procedure ("*Genehmigung*"), the applicant must be granted the approval if all required conditions are met,[149] and in a plan-approval procedure ("*Planfeststellung*") a balancing of interests must be carried out before the approval may be granted. The latter procedure gives the BSH more leeway in its decision making. For OWFs, a plan-approval procedure is required.[150]

In particular, it must be examined whether marine environmental features such as birds, fish, marine mammals, benthos, sea bottom and water are put at risk by the OWF.[151] Moreover, the project shall not constitute a hazard to navigation. Further aspects concern national or allied defense and other public law requirements. OWF-approvals expire after 25 years, but after this time an extension of approval can be granted. The construction of the turbines must start within 2.5 years after notification of the approval which will otherwise become void.

Technical instructions support operators of OWFs in terms of legal security. The BSH has issued several standard requirements such as the "Ground Investigation Standards" and the "Standards on the Design of Offshore Wind Turbines."[152] These instructions cover technical requirements for the installation and operation of OWFs as part of the approval.

The approval procedure is regulated in the SeeAnlV and the more general Administrative Procedures Act ("*Verwaltungsverfahrensgesetz – VwVfG*"). It is carried out in the following steps. Initially, the operator must hand in the official application, which is then checked by the BSH for its completeness. This is followed by a first round of participation, including the Waterways and Shipping Directorate-General, the mining authority, Federal Environmental Agency, Federal Agency for Nature Conservation which are given the opportunity to give comments on the application. Section 3 SeeAnlV provides a "competition rule" regarding later applications inconsistent with the first application. The BSH can then defer such applications. In a further participation round, a larger group of stakeholders such as nature protection NGOs, commercial and small craft shipping, fisheries and wind energy associations is invited to give comments. The coastal state is also involved because all cables in the territorial sea connecting the OWF to the national grid must be approved. In a further step, the applicant can present and discuss his project during an application conference. OWF-turbines regularly exceed a height of 50 m, making an EIA according to the EIA Act (UVPG) necessary.[153] The plan must include various information, for example on precautionary and security measures, a timetable and action plan, when appropriate, a certificate from a recognized expert on the safety requirements, and a risk analysis concerning the potential collision of vessels with the turbines. The complete documentation is then sent to the competent authorities and associations, followed by a discussion with the stakeholders. If all requirements are met, the BSH will eventually

149. Section 5 para. 6 SeeAnlV.
150. Section 2 para. 1 SeeAnlV.
151. Regarding EIA requirements, *see* section §1.03[B][2].
152. *See BSH*, Licensing procedure for offshore windfarms, 2007 et seq., http://www.bsh.de/en/Products/Books/Standard/index.jsp (accessed November 26, 2016).
153. *See* below under section §1.03[B][2].

grant the official plan approval which is published in the German notices to mariners (NfS) and in two national papers.[154]

Incidental provisions of the plan approval regularly cover requirements regarding the following aspects:

- safety in the construction phase,
- a state-of-the-art geotechnical study,
- use of state-of-the-art methods in the construction of wind turbines, prior to start-up,
- presentation of a safety and protection concept,
- installation of lights, radar, and the automatic identification system (AIS) on the turbines,
- use of environmentally compatible materials and non-glare paint,
- hull-retaining construction of the substructure of the foundations,
- noise reduction during turbine construction and low-noise operation,
- presentation of a bank guarantee covering the cost of deconstruction.[155]

The procedure may seem highly complex but, due to the large number of stakeholders involved, it gives developers and operators a high level of legal certainty. Once they are granted a plan approval, this gives them a subjective right to implement the plan. This right cannot be withdrawn easily. It is considered a property right protected by, for instance, Articles 12 and 14 Basic Law ("*Grundgesetz*") and can be defended against third parties and the competent authorities.

The license scheme for the installation of OWFs does not cover an Electricity Generation License. The promotion of OWFs via Feed-in-Tariffs is harmonized with the Offshore Development Plan and the grid connection, but there is no specific license for the generation of electricity. The EnWG only requires a license for the operation of a new grid.[156]

[2] Environmental Impact Assessment

Every OWF with more than twenty turbines requires an EIA before approval. Under certain conditions, OWFs with more than three turbines may also need an EIA.[157] Practically, every OWF-project must undergo the EIA-procedure. Under German law, the EIA is not a separated procedure but an integral part of the approval procedure as described above. The EIA-procedure consists of the main steps: scoping, documentation of impacts of the main option and of alternatives, participation of relevant authorities and the public, and taking the results into account in the final decision.

154. Section 74 para. 5 VwVfG; for a list of the approved OWFs and grid connections *see* http://www.bsh.de/en/Marine_uses/Industry/Wind_farms/index.jsp (accessed November 26, 2016).
155. *BSH*, Plan approval procedure for the construction and operation of installations in the EEZ, 2016 (http://www.bsh.de/en/Marine_uses/Industry/Wind_farms/Approval_Procedure.jsp) (accessed November 26, 2016).
156. Section 4 EnWG.
157. *See* attachment 1 No. 1.6 of the UVPG.

The BSH has issued a standard concept for the implementation of the EIA-requirements with a special focus on OWFs.[158] The concept describes framework conditions for the construction, operation and decommissioning-phase and technical instructions for features of conservation interest such as benthos, fish, avifauna (resting and migratory birds), marine mammals, bats and landscape.

[3] Strategic Environmental Assessment (SEA)

SEA is also regulated by the UVPG. Two sorts of plans relevant for OWFs require an SEA. One is the Federal Spatial Planning for the EEZ ("*Raumordnung in der deutschen Ausschließlichen Wirtschaftszone*"), the other is the Spatial Offshore Grid Plan "*Bundesfachplan Offshore*").[159] Similar to EIA, the SEA is an integral part of the main planning procedure. It consists of the usual steps: screening, scoping, environmental report, consultations and the final decision. Apart from EIA, an SEA also requires an assessment of reasonable alternatives and monitoring measures.[160] The UVPG requires that both assessments, SEA and EIA, shall be combined to avoid double work. When identifying the scope of the assessment it shall be considered at which step, either the preliminary spatial planning or the project-oriented plan approval, environmental impacts are dealt with.[161] From the developers' and operators' perspective, it could be preferable to assess as many aspects as possible in the SEA-procedure, for the simple reason that spatial planning in the EEZ is a federal obligation and paid for by the government. In practice though, the EIA-procedure is regularly paid by the developer and operator. In particular, producing the environmental report can be very costly.

[4] Planning the Legal Regime: Marine Planning

The Spatial Planning Act ("*Raumordnungsgesetz*" - ROG)[162] also covers the German EEZ, and spatial development plans have been introduced. The spatial plans comprise five spatial development principles for the EEZ, namely the safeguarding and strengthening of shipping, the strengthening of economic power in the EEZ, the promotion of offshore wind-energy use, long-term safeguarding of the potentials and characteristics of the EEZ and the safe-guarding of the natural resources.[163]

The spatial development plans should contribute to the successful implementation of the 2002 governmental strategy on offshore wind-energy use. Its objectives, such as the reduction of dependency on energy imports and the increase of its

158. *BSH*, Standard Investigation of the Impacts of Offshore Wind Turbines on the Marine Environment (StUK4), 2013 (http://www.bsh.de/en/Products/Books/Standard/7003eng.pdf) (accessed November 26, 2016).
159. *See* attachment 3 no. 1.6 and 1.14 of the UVPG; with regard to these plans, *see* the next Chapter.
160. Cf. *EU Commission*, Strategic Environmental Assessment - SEA, 2016 (http://ec.europa.eu/environment/eia/sea-legalcontext.htm) (accessed November 26, 2016).
161. Cf. section 14f para. 3 UVPG.
162. Raumordnungsgesetz of December 22, 2008, Federal Law Gazette I p. 2986.
163. *See* Anlage zur Verordnung über die Raumordnung in der deutschen ausschließlichen Wirtschaftszone in der Nordsee (AWZ Nordsee-ROV), September 21, 2009.

environmental compatibility, with a special focus on climate protection, form the basis for the spatial development plans of the EEZ. These plans address the goals according to the Federal Governments' 2007 Integrated Energy and Climate Programme ("*Integriertes Energie- und Klimaprogramm IEKP*")[164] to increase the share of renewable energy in the electricity system to at least 30% by 2020. This goal was stipulated in the EEG 2009. The designation of priority areas for wind-energy use in the EEZ aims at an increased level of planning-security for investors of offshore wind farms.[165] The EEZ-spatial development plans cannot be compared to onshore state or regional planning; in the EEZ, there are no plans at lower level such as urban land-use plans for onshore wind-energy projects. The EEZ-spatial development plans address various issues, such as Natura-2000 areas, safety of maritime traffic, fishery or national defense. According to these plans, the hub height of offshore wind turbines is limited to 125 m, insofar as these installations are visible from the coast.[166] Priority areas for OWFs in the EEZ cover app. 880 km^2 in the North Sea and app. 130 km^2 in the Baltic Sea.[167]

[5] International and European Law

The most important international treaty regarding OWFs in the EEZ is the UNCLOS of December 10, 1982.[168] In its Articles 55 et seq., it provides regulations for the EEZ, beginning with a definition of the EEZ in Article 55:

> The exclusive economic zone is an area beyond and adjacent to the territorial sea, subject to the specific legal regime established in this Part, under which the rights and jurisdiction of the coastal State and the rights and freedoms of other States are governed by the relevant provisions of this Convention.

Article 56 on rights, jurisdiction and duties of the coastal State in the EEZ grants the coastal State:

> (a) sovereign rights for the purpose of exploring and exploiting, conserving and managing the natural resources, whether living or non-living, of the waters superjacent to the seabed and of the seabed and its subsoil, and with regard to other activities for the economic exploitation and exploration of the zone, such as the production of energy from the water, currents and winds;
> (b) jurisdiction as provided for in the relevant provisions of this Convention with regard to:
>   (i) the establishment and use of artificial islands, installations and structures;

164. *See* http://www.bmwi.de/DE/Service/gesetze,did=254040.html (accessed November 26, 2016).
165. Anlage zur Verordnung über die Raumordnung in der deutschen ausschließlichen Wirtschaftszone in der Nordsee (AWZ Nordsee-ROV), September 21, 2009, 2.3 and 3.5.
166. Anlage zur Verordnung über die Raumordnung in der deutschen ausschließlichen Wirtschaftszone in der Nordsee (AWZ Nordsee-ROV), September 21, 2009, 3.5.1.
167. Anlage zur Verordnung über die Raumordnung in der deutschen ausschließlichen Wirtschaftszone in der Nordsee (AWZ Nordsee-ROV), September 21, 2009, 3.5.
168. *See United Nations*, http://www.un.org/depts/los/convention_agreements/texts/unclos/unclos_e.pdf (accessed November 26, 2016).

(ii) marine scientific research;
(iii) the protection and preservation of the marine environment;
(c) other rights and duties provided for in this Convention.

Article 56 also covers the right to install OWFs. Partly relevant to OWFs are also the UN Convention on Biological Diversity,[169] the UN Framework Convention on Climate Change – UNFCCC[170] – or the 2015 Paris Agreement.[171]

Regarding access to information, public participation and access to courts, the United Nations Economic Commission for Europe's (UNECE) Aarhus Convention is of the utmost importance.[172] The convention has been signed and ratified by Germany and the EU.

Special international treaties such as the Convention on the Protection of the Marine Environment of the Baltic Sea Area (Helsinki Convention – HELCOM)[173] and the Convention for the Protection of the Marine Environment of the North-East Atlantic ('OSPAR Convention'),[174] both from 1992, aim at protecting the maritime environment. The OSPAR-Convention, for instance, forbids "any dumping of wastes or other matter from offshore installations."[175] For the practical planning and approval procedure, however, the conventions play only a marginal role.

At EU-level, almost all national environmental laws derive from EU-law. For OWFs, for example, the Marine Strategy Framework Directive[176] determines qualitative descriptors for a good environmental status such as:

> Introduction of energy, including underwater noise, is at levels that do not adversely affect the marine environment.[177]

Further EU-laws with relevance to OWFs include the Water Framework[178] and the Habitats Directive.[179] Under the latter, Special Areas of Conservation within the

169. *See* the full text under https://www.cbd.int/convention/text/default.shtml (accessed November 26, 2016).
170. *See* the full text under *United Nations*, http://unfccc.int/files/essential_background/background_publications_htmlpdf/application/pdf/conveng.pdf (accessed November 26, 2016).
171. *See* the full text under *United Nations*, http://unfccc.int/files/essential_background/convention/application/pdf/english_paris_agreement.pdf (accessed November 26, 2016).
172. Convention on Access to Information, Public participation and Access to Courts in Environmental Matters, done at Aarhus, Denmark, on June 25, 1998, *see* the full text under *United Nations Economic Commission for Europe*, http://www.unece.org/fileadmin/DAM/env/pp/documents/cep43e.pdf (accessed November 26, 2016).
173. *See* the full text under http://www.helcom.fi/Documents/About%20us/Convention%20and%20commitments/Helsinki%20Convention/Helsinki%20Convention_July%202014.pdf (accessed November 26, 2016).
174. *See* the full text under http://www.ospar.org/site/assets/files/1290/ospar_convention_e_updated_text_in_2007_no_revs.pdf (accessed November 26, 2016).
175. Annex III, Art. 3 para. 1 OSPAR-Convention.
176. Directive 2008/56/EC of June 17, 2008 establishing a framework for community action in the field of marine environmental policy, Official Journal of the European Union L 164/19.
177. Annex I No. 11 of Directive 2008/56/EC.
178. Directive 2000/60/EC of October 23, 2000 establishing a framework for Community action in the field of water policy, Official Journal of the European Communities L 327/1.
179. Council Directive 92/43/EEC of May 21, 1992 on the conservation of natural habitats and of wild fauna and flora, Official Journal of the European Communities No L 206 / 7.

Natura 2000 network can be set up. For instance, three Natura 2000 areas were established in the North Sea, i.e., the *"Doggerbank," "Borkum-Riffgrund," "Sylter Außenriff"* and *"Östliche Deutsche Bucht."*[180] The installation of OWFs is not allowed within these protected areas.

The 2009 Renewable Energies Directive[181] provides obligations for Member States' administrative procedures in Article 13:

> 1. Member States shall ensure that any national rules concerning the authorisation, certification and licensing procedures that are applied to plants and associated transmission and distribution network infrastructures for the production of electricity, heating or cooling from renewable energy sources, and to the process of transformation of biomass into biofuels or other energy products, are proportionate and necessary.
>
> Member States shall, in particular, take the appropriate steps to ensure that:
>
> (a) subject to differences between Member States in their administrative structures and organisation, the respective responsibilities of national, regional and local administrative bodies for authorisation, certification and licensing procedures including spatial planning are clearly coordinated and defined, with transparent timetables for determining planning and building applications;
>
> (b) comprehensive information on the processing of authorisation, certification and licensing applications for renewable energy installations and on available assistance to applicants are made available at the appropriate level;
>
> (c) administrative procedures are streamlined and expedited at the appropriate administrative level;
>
> (d) rules governing authorisation, certification and licensing are objective, transparent, proportionate, do not discriminate between applicants and take fully into account the particularities of individual renewable energy technologies;
>
> (e) administrative charges paid by consumers, planners, architects, builders and equipment and system installers and suppliers are transparent and cost-related; and
>
> (f) simplified and less burdensome authorisation procedures, including through simple notification if allowed by the applicable regulatory framework, are established for smaller projects and for decentralised devices for producing energy from renewable sources, where appropriate.... .

These rules give national authorities a guideline for the handling of permits for renewable energy installations, including OWFs.

[6] Public Participation Scheme

Public participation forms an essential part of the SEA for the spatial planning procedure, and also of the EIA for the plan approval of OWFs. The underlying Aarhus

180. *See Federal Agency for Nature Conservation* (BfN), Schutzgebiete in der Nordsee (AWZ) (2016) (https://www.bfn.de/0314_nordsee_meeresschutzgebiete.html) (accessed November 26, 2016).

181. Directive 2009/28/EC of April 23, 2009 on the promotion of the use of energy from renewable sources, Official Journal of the European Union L 140/16.

Convention has been implemented by EU- and national law, in particular the SEA- and EIA-directives,[182] the UVPG and the Act on Standing in Environmental Matters ("*Umwelt-Rechtsbehelfsgesetz – UmwRG*").[183] The BSH must involve the public by announcing the plan or project, making available the documents for inspection for a reasonable period of time, giving an opportunity for the public to express its opinion, and informing the public about the decision.[184] For instance, in the SEA-procedure, the draft-plan must be made public for at least one month, and then the public concerned has at least one further month to express its opinion. A public hearing is no longer obligatory,[185] although it is still common. The Spatial Offshore Grid Plan, for instance, was discussed in a public hearing on October 30, 2012.[186] Public participation does not constitute making legal claims.[187]

[7] Grid Expansion

Grid expansion plays a crucial role for OFWs. To connect onshore- and offshore wind farms in the north to the centers of energy consumption in the west and south of Germany, several thousand kilometers of new extra-high voltage cables are needed. The Grid Expansion Acceleration Act ("*Netzausbaubeschleunigungsgesetz Übertragungsnetz" – NABEG*)[188] and the amendment of the Energy Act (EnWG) passed in 2011 form part of the legal framework for the expansion of Germany's extra-high voltage networks.

Grid expansion and grid connection of OWFs are closely related. The Spatial Offshore Grid Plan, for instance, forms a pre-condition for their approval. According to the BSH, it covers:

- the identification of offshore wind farms suitable for collective grid connections,
- spatial routing for subsea cables required to connect offshore wind farms,
- sites for converter platforms or transformer substations,
- interconnectors,
- a description of potential cross connections between grid infrastructures and

182. Directive 2001/42/EC of June 27, 2001 on the assessment of the effects of certain plans and programmes on the environment, Official Journal of the European Communities L 197/30; Directive 2011/92/EU of December 13, 2011 on the assessment of the effects of certain public and private projects on the environment (codification), Official Journal of the European Union L 26/1.
183. *See BNetzA*, Creating Future-Proof Networks, 2016 (http://www.netzausbau.de/EN/home/en.html;jsessionid=E16A7D725D5F6D049487886D37961C91) (accessed November 26, 2016).
184. Section 9 para. 1 UVPG.
185. Section 14i para. 3 UVPG.
186. *BSH*, Spatial Offshore Grid Plan for the German Exclusive Economic Zone of the North Sea and Non-technical Summary of the Environmental Report 2012, 2012, p. 11 (http://www.bsh.de/en/Marine_uses/BFO/BFO-N2012_ComprehensiveSummary_INTERNET.pdf) (accessed November 26, 2016).
187. Regarding the right of standing *see* section §1.02[C].
188. Netzausbaubeschleunigungsgesetz Übertragungsnetz of July 28, 2011, Federal Law Gazette I p. 1690.

- standardised technical rules and planning principles.[189]

Moreover, the new WindSeeG with its Spatial Development Plan ("*Flächenentwicklungsplan*") aims at an orderly and efficient utilization of OWFs grid connections. Such connections and the further expansion of offshore wind energy should proceed in a coordinated manner.[190]

§1.04 CHALLENGES AND SOLUTIONS

In the light of recent numbers of OWFs, their development over the last years seems a success: the total capacity has increased from 628.3 MW in 2014 to 3,982.2 MW in 2016, meaning 55% of this goal of the 2020 target of 6,500 MW has already been reached.[191] However, various obstacles and challenges for the further expansion of OWFs remain.

Regarding the institutional design, the situation may not be perfect, but the legal and organizational framework has improved a lot in the last years. Compared to the beginning of OWFs-licensing procedures, the BSH now has a highly competent experienced staff. Procedures are carried out relatively smoothly. One improvement is that public resistance against OWFs seems to be relatively low, making the procedures less controversial than, for instance, onshore wind farm planning and permit procedures. Despite much conflicts on details, generally speaking, the competent authorities such as the BSH, the BNetzA and the BfN work closely together.

Still, one of the major challenges is financing the OWFs. One park with eighty turbines needs between EUR 1.2 and EUR 2.0 billion. Complex financial plans are needed with a variety of equity and debt financing. Project financing can form an alternative.[192] With regard to such high financial investments, citizens' financial participation does not seem appropriate. Financial incentives are therefore of the highest importance. The legislator has reacted to such challenges. Under the current EEG 2014, compared to other sources of renewable energies, OWFs get the highest tariffs for electricity feed-in. It is not yet quite clear what changes the coming tender and auctioning scheme under the new WindSeeG will bring. It is unlikely that the prices will decrease. It is probable that OWFs will not be competitive compared to, for example, onshore wind energy, for quite some time. This might, however, change in the long run. For the moment, it seems that the German legislator has made relatively well-founded decisions which could enable the offshore wind-energy industry to become financially feasible.

---

189. *BSH*, Spatial Offshore Grid Plan (2016) (http://www.bsh.de/en/Marine_uses/BFO/index.jsp) (accessed November 26, 2016).
190. Section 4 WindSeeG.
191. *Foundation Offshore Wind Energy*, Windenergie auf See (2016) (http://www.offshore-stiftung.de/sites/offshorelink.de/files/documents/Wanderausstellung%20web_neu_August%202016.pdf) (accessed November 26, 2016).
192. *Foundation Offshore Wind Energy*, Finanzierung von Offshore-Windparks (2012) (http://www.offshore-stiftung.de/sites/offshorelink.de/files/documents/Factsheet%20Finanzierung%20von%20Offshore-Windparks%2009_2012.pdf) (accessed November 26, 2016).

The regulations for the installation, operation and remuneration of OWFs have also changed greatly in recent years. The German legislator has passed a large number of laws either directly or indirectly relevant to OWFs. The new EEG 2017, the WindSeeG, the amendments of the EnWG, the NABEG, etc. all aim at implementing the energy transition goals ("*Energiewende*"). It will take some time before the effects of these new laws can be evaluated. For the time being, one could say that the regulations are highly, if not overly, complex. These details are necessary if OWFs are to be sufficiently promoted.

§1.05 CONCLUSION

OWFs have the chance of becoming a major pillar of Germany's energy transition. They are especially necessary in the step-by-step replacement of nuclear power. All the same, the goal of a 25 GW capacity of OWFs by 2030 seems very ambitious. There is widespread insecurity as to whether the new laws concerning the promotion of OWFs will be sufficiently effective for reaching this target. The legal situation is becoming more and more complex and even specialized lawyers struggle to understand all its intricacies. The key, not only for OWFs, but also for the whole renewable energy sector is higher energy prices. Once production costs for "old" sources such as fossil (and nuclear) rise, the gap between cost-intensive technologies, such as OWFs and these old energies will become smaller, making the offshore wind-energy industry more feasible. Legal instruments for promoting OWFs are only designed to further innovation and new installations for a limited time. Eventually, the economic situation of the energy industry in general will become the most important factor.

CHAPTER 2

Legal Framework to Develop Offshore Wind Power in United Kingdom

Navraj Singh Ghaleigh[*]

§2.01 INTRODUCTION

The United Kingdom (UK) views offshore wind [OSW] as a key technology in its long-term plans to decarbonize its electricity generation. Decarbonization policy making in the UK has long operated within the framework of the energy trilemma - the challenge of generating power which is cost effective, low carbon, and guarantees security of supply.[1] In this framing, OSW provides a number of advantages over other forms of renewables. It avoids many of the problems associated with onshore wind (noise, flicker, aesthetic objections, etc.), and intermittency with higher load factors, huge wind resources, and diminished (though certainly not negligible)[2] impacts on visibility. The industrial development opportunity offered by OSW (as an export market) is also a significant factor for the UK.

As of April 2015, the UK had 5.7 GW installed capacity or under construction, with an additional 10 GW consented and to be delivered by 2020. This represents the largest expansion in any class of renewable energy technology, and means that the UK alone accounts for 41 % of the global total of OSW generating capacity. The reasons for this pre-eminence are both regulatory and physical. The North Sea (the body of water along the easterly coast of England and Scotland) provides possibly the planets optimal

* Sincere thanks to Ms Dagmar Topf Aguiar de Medeiros for her superb research assistance.

1. For example UK Parliament Energy And Climate Change Committee, "Energy and Climate Change Committee 3rd Report. The Energy Revolution and Future Challenges for UK Energy and Climate Change Policy Volume 1" <http://publicinformationonline.com/download/121292> accessed April 27, 2017.
2. Maria Lee, "Knowledge and Landscape in Wind Energy Planning" (2017) 37 Legal Studies 3; and discussion at section §2.04 below.

site for OSW turbine, both within territorial waters and within its continental shelf. Its waters are relatively shallow and calm, and in close proximity to numerous port facilities suitable for the assembly and transport of turbine towers, rotor, and blades. Installed turbines are also relatively proximate to load centers, aiding the economics of projects. Above all, the wind conditions (speed, variability, etc.) are well suited to existing technological limits.[3] As regards the regulatory environment, the UK consistently tops international rankings as leading location for OSW investment.[4] Further, the UK enjoys a reputation for operating stable and predictable policy regimes to support investment in renewable electricity infrastructure – discussed below. At the time of writing, OSW generates c. 6% of the UK's annual electricity requirements, a figure which is expected to grow to 10% by 2020.

[A] The Emission Reduction and/or Renewable Energy Duty/Target of the Countries at Stake

The UK has committed itself through the Climate Change Act (CCA) 2008[5] ('the Act') to a legally binding commitment of an emissions reduction of at least 80% by 2050, against a 1990 baseline.[6] This target is considered to represent an appropriate contribution to global emission reductions consistent with limiting global temperature rises to 2°C.

To ensure that regular progress is made towards the long-term target, the Act also establishes a series of five-yearly carbon budgets. The first four carbon budgets, leading to 2027, have been set,[7] and the UK is currently in the second of these carbon budget periods (2013–2017). Meeting the fourth carbon budget (2023–2027) will require that emissions be reduced by 50% on 1990 levels in 2025.[8] The Climate Change Committee (the key, independent advisory body[9]) published its advice to the Government on the fifth carbon budget in November 2015, covering the period 2028–2032, as required under section 4 of the CCA. The Government has accepted that advice in October 2016 (albeit that the UK Government rejected the formal inclusion of emissions from international shipping within the fifth carbon budget, contrary to the recommendation

3. *See* generally, https://www.thecrownestate.co.uk/energy-minerals-and-infrastructure/offshore-wind-energy/ (accessed April 27, 2017).
4. That is, Ernst and Young Country Attractiveness Index Offshore Wind (2015).
5. Climate Change Act 2008 http://www.legislation.gov.uk/ukpga/2008/27/pdfs/ukpga_20080027_en.pdf.
6. Climate Change Act 2008 section 1(1).
7. The first three carbon budgets were established in Statutory Instruments 2009 No. 1259 Climate Change The Carbon Budgets Order 2009 May 20, 2009 http://www.legislation.gov.uk/uksi/2009/1259/pdfs/uksi_20091259_en.pdf. The fourth carbon budget was established in Statutory Instruments 2011 No. 1603 The Carbon Budget Order 2011 June 29, 2011 http://www.legislation.gov.uk/uksi/2011/1603/pdfs/uksi_20111603_en.pdf.
8. Although the UK has outperformed its first target and is set to outperform on the second and third targets as well, the Climate Change Committee predicts that meeting the fourth budget will not be feasible unless the government applies more challenging measures. https://www.theccc.org.uk/tackling-climate-change/reducing-carbon-emissions/carbon-budgets-and-targets/.
9. *See* Climate Change Act Part 2 and Schedule 1.

of the Committee on Climate Change (CCC)) and will legislate to that effect in the short term.[10]

The CCC recommended that the UK reduce its greenhouse gas emissions in 2030 by 57% relative to 1990 levels.[11] *See* below for details:

Budget	*Carbon Budget Level*	*% Reduction Below Base Year*
1st Carbon budget (2008–2012)	3,018 MtCO2e	23%
2nd Carbon budget (2013–2017)	2,782 MtCO2e	29%
3rd Carbon budget (2018–2022)	2,544 MtCO2e	35% by 2020
4th Carbon budget (2023–2027)	1,950 MtCO2e	50% by 2025
5th Carbon budget (2028–2032)	1,765 MtCO2e	57% by 2030

Whilst EU law, and in particular the Climate and Energy Package 2009,[12] forms a significant contextual factor for UK renewables law and targets, this chapter focuses on UK law. Furthermore, it remains to be seen what the importance of EU law will be post-Brexit. The vote to leave the EU does not in and of itself change the UK's legal commitments to reduce its emissions by 57% by 2030 and at least 80% by 2050 (relative to 1990) under the CCA.

Meeting the UK's domestic 2030 target and complying with its commitments under the Paris Agreement will also require new and renewed policies through the 2020s. Following the vote to leave the EU, the Government will also need to preserve, replicate, or take opportunities to improve on areas of policy previously agreed by the UK at the EU level.[13] As they stand today, existing policies would at best deliver around half of the required emissions reduction to 2030. The Government has committed to closing that gap in its forthcoming Emissions Reduction Plan.[14]

[B] Why Move from Onshore to Offshore?

The UK's engagement with offshore wind stems from a number of factors.

10. *See* generally, https://www.gov.uk/government/uploads/system/uploads/attachment_data/file/559954/57204_Unnumbered_Gov_Response_Web_Accessible.pdf.
11. The fifth carbon budget was established in Statutory Instruments 2016 No. 785 The Carbon Budget Order 2016 July 20, 2016 http://www.legislation.gov.uk/uksi/2016/785/pdfs/uksi_20160785_en.pdf.
12. *See* generally, http://ec.europa.eu/clima/policies/strategies/2020/index_en.htm.
13. https://www.theccc.org.uk/publication/meeting-carbon-budgets-implications-of-brexit-for-uk-climate-policy/.
14. https://www.theccc.org.uk/2016/10/13/concrete-action-needed-to-meet-uk-climate-commitments-following-paris-agreement-and-brexit-vote/.

Geographically, offshore wind is an attractive option, because, as an island nation with coasts on the Atlantic Ocean and North Sea, the UK has very significant wind resources.

From an economic policy point of view, offshore wind is seen as an opportunity for the UK to be a "key part of a global supply chain that delivers and sustains jobs, exports and economic benefits for the UK."[15]

Additionally, from the perspective of the wider public, offshore is an attractive option because there is a widespread perception (especially in the more Conservative Party-leaning parts of rural England) that onshore wind is a blot on the landscape and as well as being unsightly, causes problems of noise and flicker for local communities.

All these considerations combined add to the attraction of including offshore wind as a part of the UK's long-term, low carbon, electricity mix. The increasing popularity of offshore wind is demonstrated by the fact that installed capacity in 2016 was 5.1 GW with a further 4.5 GW under construction, representing c. 8%–10% of the UK's electricity supply by 2020.[16] The increased interest in offshore wind energy is furthermore demonstrated by the fact that the latest figures show that the current load factor for offshore is 36.9% compared with 27.3% for onshore wind.[17]

The full market potential of the UK's offshore wind resources was published in May 2010 by the Offshore Valuation Group, an informal collaboration of government and industry organizations.[18] Their findings revealed that, using less than a third of the total available resource, the offshore renewable energy industry in the UK could generate electricity equivalent to 1 billion barrels of oils annually (this would match North Sea oil and gas production) and result in cumulative carbon dioxide savings of 1.1 billion tonnes by 2050.[19] The load factor for offshore is 36.9% compared with 27.3% for onshore wind.[20]

[C] Policy Setting

The policy literature on offshore wind power is voluminous. At its core is the desire to develop cost-effective programs. Key elements in achieving cost effectiveness are:

- developing clearly outlined government policies that provide sufficient long-term security to ensure investor confidence;
- increased government investment in offshore wind to secure long-term competitiveness of this type of energy;

15. Her Majesty's Government, *Offshore Wind Industrial Strategy: Business and Government Action* (HM Government 2013).
16. http://www.thecrownestate.co.uk/energy-minerals-and-infrastructure/offshore-wind-energy/.
17. http://www.renewableuk.com/page/UKWEDExplained.
18. http://publicinterest.org.uk/offshore/achnowledgements.php lists all the participating organizations.
19. http://publicinterest.org.uk/offshore/.
20. http://www.renewableuk.com/page/UKWEDExplained.

- development of government policy with regard to what happens to offshore installations after their initial contracts expire. Depending on the installation, this might be renewal of contract or repowering of the site.

A representative sample of policy programs and targets includes the following:

- *Overarching National Policy Statement for Energy EN-1 (2011)*[21] – this sets out the national policy for the energy infrastructure.
- *Contracts for Difference Policy Paper March 13, 2017.*[22] A Contract for Difference (CfD) is a private law contract between a low-carbon electricity generator (i.e., OSW developer) and the Low Carbon Contracts Company (LCCC), a government-owned company, introduced as part of the now implemented Electricity Market Reform (EMR) program. A generator party to a CfD is paid the difference between the "strike price" – a price for electricity reflecting the cost of investing in a particular low-carbon technology – and the "reference price"– a measure of the average market price for electricity in the GB market. CfD are a critical element of the UK low-carbon architecture, seeking to provide greater certainty and stability of revenues to electricity generators by reducing their exposure to volatile wholesale prices, whilst protecting consumers from paying for higher support costs when electricity prices are high.
- Renewable Energy Review (May 2011). The Review was published by the Climate Change Committee, in accordance with its instruction from the government in the May Coalition Agreement 2010.[23] In the Review, the Climate Change Committee recommends a portfolio approach that includes a mix of low-carbon generation technologies such as onshore, offshore, nuclear, and carbon capture storage. Although offshore is currently one of the more expensive technologies, its capacity (especially when taken into consideration the local landscape impact of onshore) and its potential for cost reduction[24] make it a sensible policy choice.[25]

The fifth carbon budget – The next step towards a low-carbon economy (CCC, October 22, 2015).[26] *Inter alia*, this points out the importance of the low-carbon portfolio including roll-out in the 2020s of offshore wind and carbon capture and storage (CCS) given their long-term importance and the role of UK deployment in driving down costs.

21. https://www.gov.uk/government/uploads/system/uploads/attachment_data/file/47854/1938-overarching-nps-for-energy-en1.pdf.
22. https://www.gov.uk/government/publications/contracts-for-difference/contract-for-difference.
23. Executive Summary of the Renewable Energy Review, p. 10.
24. For more information on cost reduction opportunities *see*: Approaches to cost-reduction in offshore wind: A report for the Committee on Climate Change (BVG, June 2015).
25. Executive Summary of the Renewable Energy Review, p. 12.
26. http://tinyurl.com/z83agnf.

Offshore wind: A valuable ingredient in the UK's future energy mix (CCC, December 2, 2015).[27] This recommends that the UK Government extends funding under the Levy Control Framework to at least 2025 and continues to offer low-carbon contracts via auction. Within this, offshore wind should be supported until it is cost competitive with a new build unabated gas plant by setting out an intention to contract 1–2 GW per year, provided costs fall, and with a view to removing subsidy in the 2020s.

§2.02 INSTITUTIONAL DESIGN FOR OFFSHORE WIND

[A] The "Main" Authority During the Development of Offshore Wind Power

Until July 2016, the primary authority responsible for UK offshore wind policy was the Department of Energy and Climate Change (DECC). However, with the new government of Prime Minister Teresa May, the Department of Energy & Climate Change became part of Department for Business, Energy & Industrial Strategy (BEIS) in July 2016.[28] BEIS is in charge of energy policy in general and as such is the main authority in drafting and preparing policy and law for the energy sector, including offshore energy.

As the name suggests, the new responsible department is primarily an energy and industry authority, rather than a climate or environmental one. This has raised concerns that climate change would be placed further down the agenda, to which the Government has responded that climate change will be part of BEIS' brief, and that climate change will be governed by an enlarged and more influential department.

It should nonetheless the noted that even prior to this departmental reorganization, key aspects of renewables policy (i.e., support mechanisms such as the EMR) were jointly steered by a triumvirate of departments, namely DECC, BIS, and UKTI.[29] These departments, albeit in their new formats, continue to play a key role in the development of offshore wind projects. BEIS is responsible for energy policy in general and as such is the main authority in drafting and preparing policy and law for the energy sector, including offshore energy.

[B] The "Supplementary" Authorities During the Development of Offshore Wind Power

[1] Central Government Level

R&D Authority

Manufacturing Advisory Service. This was a creation of the Department for Business, Innovation & Skills (BIS), was split into regions and aimed at SMEs to offer technical

27. http://tinyurl.com/h7fh3uq.
28. http://tinyurl.com/zyjhnu7.
29. Under the new government of Prime Minister Teresa May, UKTI has become the Department for International Trade (DIT).

and strategic advice. It was however formally closing on March 31, 2016, as part of the Conservative government's austerity program.

During its operation it ran programs such as "GROW: Offshore Wind service," a GBP 20 million fund from the Regional Growth Fund to enable manufacturers to take advantage of the rapidly growing offshore wind market.

Similar initiatives included the Offshore Wind Investment Organisation,[30] to increase the levels of inward investment to the UK in the offshore wind supply chain, the GBP 46 Offshore Renewable Energy Catapult to drive innovation, and enabling innovative companies to commercialize their products, thereby bringing down costs for consumers.

Spatial Planning Authority

Planning permission of various sorts is required to construct a wind farm and any associated infrastructure. The planning regimes (though not substantially) differ as between England, Wales, Scotland, and Northern Ireland - the focus herein is on the English regime.

For Larger Offshore Projects

Offshore projects of more than 100 megavolts (MW) are classed as nationally significant infrastructure projects (NSIPs) and require a "development consent order" from the Secretary of State for Communities and Local Government, under the *Planning Act 2008*.[31] Sections 14 and 15 provide that larger than 100 MW is NSIP 15(1) & 15(3) and 14(1)(a). Section 31 states that development consent order is required for NSIP. Section 32 refers to TCAP 1990 for meaning of development.

Smaller Offshore Projects

For smaller offshore plant i.e., between 1 MW and 100 MW, consent is required pursuant to section 36 of the Electricity Act 1989 (known as a "section 36 consent").[32] The Marine Management Organisation (MMO) is the decision-making body.[33] The MMO is responsible for licensing offshore generating stations including wind farms, wave and tidal devices with a capacity between 1 and 100 MW. Generating stations of a greater capacity qualify as NSIPs and are subject to *a different system of consent*. In addition to a marine license, these applications may also require consent under the Electricity Act 1989. The MMO carries out licensing and enforcement functions under the Marine and Coastal Access Act (MCAA) Part 4, on behalf of the Secretary of State

30. https://www.gov.uk/government/groups/offshore-wind-investment-organisation.
31. https://www.gov.uk/guidance/offshore-wind-part-of-the-uks-energy-mix.
32. https://www.gov.uk/guidance/consents-and-planning-applications-for-national-energy-infrastructure-projects.
33. http://webarchive.nationalarchives.gov.uk/20140108121958/http://www.marinemanagement.org.uk/licensing/marine/activities/construction.htm.

in the English inshore region and all English, Welsh and Northern Ireland offshore regions.[34]

Marine Authority

The "Crown Estate"[35] owns the seabed out to the 12 nautical mile territorial limit and grants licenses for the generation of renewable energy on the continental shelf within the *renewable energy zone* (REZ) out to 200 nautical miles.

The Crown owns or has vested in it the following: over 50% of the foreshore of the UK (i.e., the area between mean low water and mean high water on the coast and tidal waters); almost the entirety of the seabed to 12 nautical miles; manages the Marine Estate on behalf of the Crown, maintaining and enhancing its value and the return obtained from it (*see Crown Estate Act 1961*), on a commercial basis; and manages the rights to generate electricity from wind, waves, and the tides on the continental shelf under the *Energy Act 2004*.

[2] The Role of the Local Government

No direct role in offshore wind. Local planning authorities are responsible for renewable and low-carbon energy development of 50 MW or less installed capacity (under the Town and Country Planning Act 1990). Renewable and low-carbon development over 50 MW capacity are currently considered by the Secretary of State for Energy under the Planning Act 2008, and the local planning authority is a statutory consultee. It is the Government's intention to amend legislation so that all applications for onshore wind energy development are handled by local planning authorities.

[3] Other Players Vital for the Development of Offshore Wind

- The Civil Aviation Authority (CAA) ensures that the construction of wind turbines does not impact on aviation in the UK. The CAA represents civil aviation interests on the DECC-chaired Aviation Management Board.
- Ofgem,[36] the energy regulator, grants and monitors compliance with the *generation license*.
- National Grid Electricity Transmission (National Grid) is the system operator of the electricity transmission system, into whose transmission or distribution network the power plant will be connected. National Grid is also responsible for the allocation of Contracts for Difference.

34. http://webarchive.nationalarchives.gov.uk/20140108121958/http://www.marinemanagement.org.uk/licensing/documents/guidance/02.pdf.
35. https://www.thecrownestate.co.uk/energy-minerals-and-infrastructure/offshore-wind-energy/.
36. https://www.ofgem.gov.uk/electricity/transmission-networks/offshore-transmission.

- The LCCC is a private company, owned by BEIS. LCCC manages the new CfD introduced by the Government as part of the ambitious EMR program.[37]
- National Grid has been appointed as the Delivery Body for Contracts for Difference, responsible for publishing CfD application/allocation guidelines and running the CfD allocation process.

[C] The Role of "Law" under the Institution Decision-Making and Related Policies

Law is essential in shaping all the four stages of offshore wind projects: consenting phase, site investigation, construction phase, and operational phase. As discussed elsewhere in this chapter, the UK has developed a sophisticated and largely effective regime for offshore wind that seeks to develop offshore installed capacity, safeguard the environment, and drive innovation in the sector.

The UK can credibly claim to be the first country in the world to introduce a long-term legally binding framework nationally to tackle the dangers of climate change, through the CCA. The Act provides the UK with a legal framework, including a long-term target for emissions in 2050, five-year carbon budgets on track to that target, and the development of a climate change adaptation plan. There are also a variety of legal regulations in place to ensure we meet our climate change obligations. The CCA (2008) is part of the government's plan to reduce greenhouse gas emissions. It established the framework to develop a targeted and economically credible plan to reduce current and future emissions.[38]

By way of the CCA, the government has set targets and created several groups and programs to work on limiting and adapting to climate change. The key institution created by the CCA is the CCC. Its function is to advise the Government on emissions targets and report to the Parliament on progress made in reducing greenhouse gas emissions. Further, the CCA commits successive UK governments to reducing greenhouse gas emissions by at least 80% of 1990 levels by 2050. This includes reducing emissions from the devolved administrations (Scotland, Wales, and Northern Ireland), which currently account for about 20% of the UK's emissions. This target was based on advice from the CCC's 2008 report, "Building a low-carbon economy."[39] Although preventing dangerous climate change, and preparing for it, touches on all aspects of the economy, and several government departments provide input into climate change policies, the two main government departments responsible are Department for BEIS[40]

37. https://www.gov.uk/government/publications/contracts-for-difference/contract-for-difference.
38. *See* generally, Mark Stallworthy, "Legislating against Climate Change: A UK Perspective on a Sisyphean Challenge" (2009) 72 The Modern Law Review 412; Matthew Lockwood, "The Political Sustainability of Climate Policy: The Case of the UK Climate Change Act" (2013) 23 Global Environmental Change 1339.
39. https://www.theccc.org.uk/publication/building-a-low-carbon-economy-the-uks-contribution-to-tackling-climate-change-2/.
40. https://www.gov.uk/government/organisations/department-for-business-energy-and-industrial-strategy.

– leading on the policy for mitigation, – and Department for Environment and Rural Affairs (Defra)[41] – leading on the adaptation policy. Additionally, BEIS is responsible for ensuring secure energy and promoting action on climate change in the UK and internationally.

The governments and assemblies of the devolved administrations (Scotland, Wales, and Northern Ireland) create climate change policy for their devolved area, and help to implement UK-wide policies. As well as being covered by the CCA, Scotland, Wales, and Northern Ireland have their own climate change policies. For example, the Climate Change (Scotland) Act 2009 commits Scotland to a 42 % reduction in emissions by 2020 and annual reductions between 2010 and 2050; the Northern Ireland Assembly is developing plans for a Northern Ireland CCA, following advice from the CCC; the Welsh government is considering how to set its own carbon budgets.

§2.03 LEGAL DESIGN FOR OFFSHORE WIND[42]

[A] Incentives

[1] Main Finance Scheme

Offshore wind farm projects in the UK are usually financed on a project finance basis (i.e., based upon the projected cash flows of the project), although it is not unknown for power producers to finance projects using their own balance sheet.

Electricity generated by the wind farm is typically sold under a long-term agreement to secure the revenue to the project for the duration of the financing term – the Power Purchase Agreement.

The UK started a scheme known as "contracts for difference" in 2015, replacing the "Renewables Obligation" scheme, as part of EMR which seeks to further stimulate low-carbon generation. Buyers under PPAs are known as offtakers, and are typically electricity suppliers or energy trading businesses although they may include large corporates, industrial customers and even public authorities contracting under "direct PPAs."

A CfD is a long-term contract between a low-carbon electricity generator and the CfD Counterparty (the LCCC Limited,[43] a limited company owned by the government), which is set at a fixed price for the electricity produced, known as the strike price. The generator sells the electricity in the wholesale market. When the market price is:

41. https://www.gov.uk/government/organisations/department-for-environment-food-rural-affairs.
42. This section draws heavily on Henry Davey and others, *Anatomy of a Wind Power Project* (Practical Law Company 2016).
43. https://www.lowcarboncontracts.uk/.

- Below the strike price, the generator receives a top-up payment from the CfD Counterparty for the additional amount (recovered from suppliers and reflected in consumers' electricity bills).
- Above the strike price, the generator must pay back the difference to the CfD Counterparty.

The CfD started operating in 2015 and will gradually replace the Renewables Obligation (RO) as financial support for new low-carbon electricity generation, which includes renewables, nuclear and CCS. EMR is aimed at large-scale low-carbon generation, so does not affect small-scale installations supported by feed-in tariffs (FITs). The first CfDs were awarded in February 2015, and there have been successive CfD Allocation Rounds.[44] The general principles behind the CfD mechanism are illustrated in the diagram below, which shows how the generator receives a top-up, or makes a payment, by reference to the strike price.

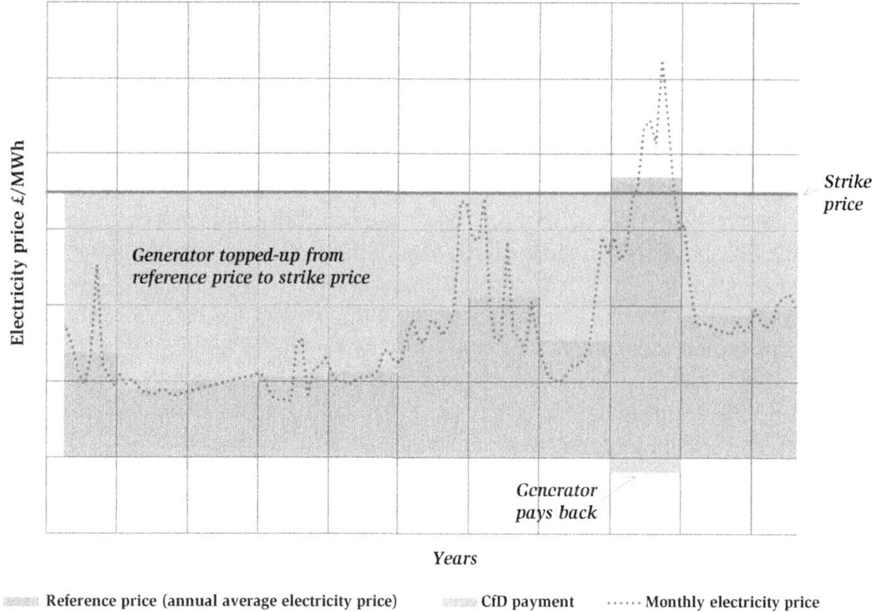

Source: This diagram is reproduced under the terms of the Click-Use Licence from DECC: Electricity Market Reform: Policy Overview (May 2012) (http://www.decc.gov.uk/assets/decc/11/policy-legislation/EMR/5349-electricity-market-reform-policy-overview.pdf).

To be eligible for a CfD, the wind farm cannot be accredited for the RO, or be subject to an NFFO PPA or a *capacity agreement*. The project company will need to ensure that any PPA that it enters into takes account of the terms of the CfD.

44. https://www.lowcarboncontracts.uk/cfd-allocation-round.

It should be noted that although wind farms are technically eligible to participate in the Capacity Market,[45] in practice it is extremely unlikely that any wind farm project will do so. The key reason is that, as an intermittent technology, wind power is an inherently unsuitable candidate for a capacity agreement. Indeed, participation is almost inconceivable without considerable technological developments (e.g., advancements in energy storage technologies might make it possible in the future). Such developments are not expected to be made in the short term.

As a consequence of EMR (and CfD) earlier methods of financing renewables such as Renewables Obligation Certificates (ROCs) and Feed-in tariffs (FITs) are being phased out.

[2] Supplementary Finance Scheme

[a] A. Low Interest Loan and Loan Guarantee

N/A

[b] Tax Incentives

The UK's ECA scheme is designed to facilitate business investment in energy-saving plant or machinery that might otherwise be too expensive. The first year allowances let businesses set 100% of the cost of the assets against taxable profits in a single tax year. This means the company can write off the cost of the new plant or machinery against the business's taxable profits in the financial year the purchase was made.[46] However, expenditure on wind turbine assets does not fit within the criteria of "environmentally beneficial plant and machinery" (as wind turbines are energy generating rather than energy saving), meaning that such expenditure cannot automatically benefit from 100% writing down allowances as enhanced capital allowances. That said, certain component parts (e.g., motors) may be eligible.

In a similar fashion, under the annual investment allowance (AIA) firms can claim capital allowances when buying plant and machinery asset, and can deduct some or all of the value of the item from their profits before you pay tax. The *AIA* is worth GBP 200,000 from January 1, 2016, and while ECAs cannot be claimed as well as the AIA on the same expenditure, the taxpayer may choose which one to claim. To the extent not eligible for ECAs, the cost of the wind turbine assets should qualify as plant and machinery with the associated capital allowances claimable. Other project costs

45. As the security of future electricity supply is uncertain, the government has decided to introduce capacity agreements within a Capacity Market to provide an "insurance policy" to reduce the likelihood of future blackouts. The capacity agreements will provide payments for reliable electricity generating capacity to be available when needed, helping to ensure security of supply.
46. https://www.gov.uk/guidance/energy-technology-list#enhanced-capital-allowance-eca-scheme.

such as professional costs and site preliminaries may also qualify for capital allowances, if it can be demonstrated that such costs are incidental to the installation or acquisition of the plant and machinery.

[c] Investment Subsidy/Grant

Regional and national grants are available from time to time in the wind farm sector. These vary according to government policy and are in line with government spending reviews. In addition to grants, there are various programs, initiatives, and organizations that offer different levels of support to the sector covering areas from supply chain, innovation, finance, and skills, with a special emphasis on research and development. For example, DECC announced significant grant funding for wind energy in 2010, and further funding as part of the Energy Entrepreneurs Fund in 2012. Examples:

Energy Entrepreneurs Fund[47]

DECC's Energy Entrepreneurs Fund is a GBP 40 million program run in phases since 2012 to support the development and demonstration of innovative, commercially viable technologies which could make a significant impact on the UK's energy security and climate objectives. The program is particularly targeted at start-up entrepreneurs and SMEs. DECC launched the fourth phase of the competition in November 2014, under which GBP 5 million was available for projects to be completed by March 2016.

SMART Scotland:[48] a discretionary grant that supports Scotland based SMEs to undertake technical feasibility studies and research and development (R&D) projects with a commercial endpoint. The fund aims to mitigate the technical risks and challenges associated with defining and developing new technologies.

Technology Strategy Board (TSB)[49]

The TSB is the government's innovation agency and helps UK companies to develop new technologies and products in order to stimulate economic growth in a number of sectors including energy. The TSB provides the core funding for the Offshore Renewable Energy Catapult center and also allocates grant funding to businesses directly via competitions. Key initiatives are:

The TSB energy program, which seeks to help develop a UK industrial base in the offshore wind sector which uses new technology development and technology transfer from parallel industries such as oil and gas as points of entry and disruptors of existing supply chains. Sixteen feasibility studies were funded in 2013 looking to introduce technology from other sectors and develop the offshore wind supply chain. It was

47. https://www.gov.uk/government/publications/energy-entrepreneurs-fund-phase-5.
48. https://www.scottish-enterprise.com/services/develop-new-products-and-services/smart-scotland/overview.
49. https://www.gov.uk/government/organisations/technology-strategy-board.

expected that TSB will provide around GBP 5 million per year in innovation grants for offshore renewables to 2015–2016.

Assisted Areas Status:[50] Increased aid intensity is available for energy and environmental investments in assisted areas. In an assisted area, investments in green technologies, research, development and innovation or encouraging a low-carbon economy can receive higher levels of aid through a regional bonus.

[d] The Links to ETS, CDM, and JI

Government can of course incentivize low-carbon generation by measures that make it more expensive to operate, or limit the permitted output from, a fossil-fuelled power plant, such as a carbon price floor or ETS. The UK's involvement in the EU ETS will be under review as a result of the UK's plans to withdraw from the EU (Brexit).

[B] Regulations

[1] License Scheme

In order for generators to export their electricity to the local distribution network, they must enter into a connection agreement with the local Distribution Network Operators (DNO). The DNO has a duty to offer terms of connection under section 16 of the Electricity Act 1989[51] and the terms of its distribution license.

The National Grid is required, under the terms of its electricity transmission license, to offer terms of connection to anyone wishing to connect to the electricity transmission system.[52] A generation license is required under *section 6(1)(a)* of the Electricity Act 1989 to generate electricity, unless one of the class exemptions applies or an individual exemption is granted by DECC. The class exemptions can apply to small generators (under 10 MW) or a medium size generator (under 100 MW) that is also supplying on site load (e.g., a manufacturing plant). License applications are made to Ofgem, who will publish details of the license application. The license will include standard conditions and may include additional terms called "special conditions." Copies of licenses and the standard conditions can be found on the e-public register on Ofgem's website.[53]

[2] Environmental Impact Assessment (EIA)

When processing development applications, BEIS considers the environmental consequences of proposals, applying European requirements for Environmental Impact

50. https://www.gov.uk/government/uploads/system/uploads/attachment_data/file/365657/BIS-14-1152-An-introduction-to-assisted-areas.pdf.
51. http://www.legislation.gov.uk/ukpga/1989/29/contents.
52. *See* generally, https://www.gov.uk/guidance/electricity-network-delivery-and-access#offshore-electricity-networks-development.
53. Davey and others (n. 42).

Assessments (EIAs).[54] EIAs identify and assess the potential seriousness of expected impacts to the environment arising from the construction, operation, and eventual decommissioning of a project. Where an EIA is required, the decision-maker's consideration of the relevant environmental information is a precondition for section 36 consent, marine license or DCO being validly issued.[55] As a function of EU law, the future of the EIA regime may be reformed after the withdrawal from the EU. EIAs are not required for small onshore projects, i.e., those involving a small number of turbines under than 15 m in height or where projects are not likely to have any significant environmental impact.[56]

The threshold for EIAs is as follows:

Project Type	*Applicable EIA Regulations*
Offshore wind farms between 1 MW and 100 MW. Overhead electric line of 132 kilovolts (kV) or more or which is to be installed in a sensitive area (except those of 220 kV or above or more than 15 km in length for which EIA is always required).	*Electricity Works (EIA) (England and Wales) Regulations 2000 (SI 2000/1927)*
Offshore wind farms (Scottish inshore region)	*Marine Works (EIA) Regulations 2007 (SI 2007/1518)*

[3] Strategic Environmental Assessment (SEA) and Siting Related Planning

A wide variety of factors have bearing on siting decision, including the availability of grid connections, site proximity to areas of outstanding natural beauty (AONB) or a site of special scientific interest (SSSI) which will prevent construction of wind turbines, and the prospect of securing access and land use when a site is privately owned or owned by the Crown Estate. Further, as analyzed in detail by Lee,[57] the impact on the development on the landscape from an environmental and practical perspective can be pivotal. Likewise the impact on local traffic and infrastructure concerns generated during the construction and decommissioning phases.

[4] Planning the Legal Regime: Marine Planning and/or Land Planning

The Planning Act (as amended by the Localism Act 2011) was introduced to streamline the decision-making process for NSIPs. The Planning Act passed responsibility for

54. https://www.gov.uk/guidance/consents-and-planning-applications-for-national-energy-infrastructure-projects#offshore-generating-stations-safety-zones.
55. *See* generally, https://www.gov.uk/government/uploads/system/uploads/attachment_data/file/43575/Electricity_Works_Regulations_-_guidance.pdf.
56. *Ibid.*
57. Lee (n. 2).

dealing with development consent applications for NSIPs to a public body called the Planning Inspectorate. The Planning Act has six stages.[58] These are:

Pre-application Stage

The process begins when the Planning Inspectorate is informed by a developer that they intend to submit an application. Before submitting an application, the developer is required to carry out extensive consultation on their proposals.

Acceptance

The acceptance stage begins when a developer submits a formal application for development consent to the Planning Inspectorate. There follows a period of up to twenty-eight days for the Planning Inspectorate, on behalf of the Secretary of State, to decide whether or not the application meets the standards required to be formally accepted for examination.

Pre-examination

At this stage, the public will be able to register with the Planning Inspectorate and provide a summary of their views of the application in writing. At pre-examination stage, everyone who has registered and made a relevant representation will be invited to attend a preliminary meeting run and chaired by an Inspector. This stage of the process takes approximately three months from the developer's formal notification and publicity of an accepted application.

Examination

The Planning Inspectorate has six months to carry out the examination. During this stage, people, who have registered to have their say, are invited to provide more details of their views in writing, which are considered by the Examining Authority.

Decision

The Planning Inspectorate must prepare a report on the application to the relevant Secretary of State, including a recommendation, within three months of the six-month examination period. The Secretary of State then has a further three months to make the decision on whether to grant or refuse development consent.

58. Details at http://infrastructure.planningportal.gov.uk/. Guidance for developers at http://infrastructure.planningportal.gov.uk/wp-content/uploads/2012/06/Advice-note-6-version-5.pdf.

Post Decision

Once a decision has been issued by the Secretary of State, there is a six-week period in which the decision may be challenged in the High Court by the administrative law procedure, judicial review.[59]

Additional planning procedures for OSW include applying for safety zones around the offshore installations. These have their own dedicated regime, the Electricity (Offshore Generating Stations) (Safety Zones) (Application Procedures and Control of Access) Regulations 2007,[60] and attendant guidance.[61]

A common reason for refusal of wind farm projects is that the proposal would have had unacceptable adverse visual impacts on the local environment. A developer must therefore conduct a rigorous site selection process that aims to identify a site location that would result in minimal adverse visual impacts. For an offshore wind project, the site selection process is constrained by the Crown Estate leasing arrangements. This was notable in the decision of the Secretary of State on September 11, 2015 to refuse development consent for the proposed Navitus Bay wind park, a large offshore wind farm (*see DECC: Planning decision for Navitus Bay offshore wind park (11 September 2015)*). The site location as defined by the Crown Estate lease granted to the developer meant that any wind farm development would be visible from the nearby UNESCO World Heritage Site, and the Secretary of State refused the consent largely on this visual impact basis.[62]

The UK has ratified the Aarhus Convention, a key international law instrument in the development of environmental effects and informational participation. *Inter alia*, it recognizes that "improved access to information and public participation in decision-making enhance the quality and the implementation of decisions."[63] As a matter of EU law, the EIA Directive[64] is the key instrument. The EIA Directive is transposed into domestic UK law by the Electricity Works EIA Regulations.

At the heart of the EIA Directive is Article 6's requirement that:

> Member States shall take the measures necessary to ensure that the authorities likely to be concerned by the project by reason of their specific environmental responsibilities are given an opportunity to express their opinion on the information supplied by the developer and on the request for development consent. To that end, Member States shall designate the authorities to be consulted, either in general terms or on a case-by-case basis. The information gathered pursuant to Article 5 shall be forwarded to those authorities. Detailed arrangements for consultation shall be laid down by the Member States.

59. For fuller details, *see* https://infrastructure.planninginspectorate.gov.uk/wp-content/uploads/2013/04/Advice-note-8-1v4.pdf.
60. http://www.legislation.gov.uk/uksi/2007/1948/pdfs/uksi_20071948_en.pdf.
61. https://www.gov.uk/government/uploads/system/uploads/attachment_data/file/80785/safety_zones.pdf.
62. Davey and others (n. 42).
63. Aarhus Convention on Access to Information, Public Participation in Decision-Making and Access to Justice in Environmental Matters (1998), Preamble.
64. 2011/92/EU.

The recent case of *RSPB v. Inchcape* ventilated many of these issues.[65] This concerned the proposal for four offshore wind farms in the Firth of Forth and Tay (Inch Cape, Seagreen Alpha, Seagreen Bravo, and Neart na Gaoithe) with a combined capacity of 2.248 GW. Consents were granted by Scottish Government on October 10, 2014 for construction of farms further to section 36 of the Energy Act 1989 and the Marine Works (EIA) Regulations 2007 reg. 22. The consents were challenged by the leading birdlife charity, the Royal Society for the Protection of Birds (RSPB). They claimed that the proposed developments posed a severe threat to a number of bird species which have protected habitats in the vicinity of the windfarms. The risks included collision with rotor blades, displacement from foraging areas with projected development sites, "barrier effects" to flight, including foraging to and from breeding colonies with consequences for adult survival, nest attendance, and chick provisioning.

Accordingly the RSPB brought actions for judicial review (a procedure by which decisions of public authorities can be challenged on specified grounds, such as procedural irregularity, or illegality) on the basis that the Scottish ministers had:

(1) failed to comply with their duties under the Electricity Works Regulations;
(2) failed to consult on environmental information;
(3) adopted an incorrect methodology as regards ornithological risk assessment; and
(4) were in breach of EU law by refusing/delaying to classify the relevant sea area as a marine special protection area.

The presiding judge of the Court of Session of the Outer House, Lord Stewart, agreed that material information had not been shared with the RSPB, and that it has been used in reaching the decision. "[this flaw] is not simply a procedural irregularity: it is substantive; and the decision is *ultra vires*....If the decision is not substantively flawed and ultra vires, the procedural irregularity is of such seriousness that the decision has to be reduced because substantial prejudice cannot be ruled out."

Of particular importance was the obligation to make relevant information available, and how to interpret the terms of the Aarhus Convention, Article 6(2), that it is a continuing obligation to provide information. In particular, "... the issuance of new reports and advice to the public authority should trigger an additional obligation to notify the public concerned. The obligation to update information is also found in the lead to this subparagraph [*of Article 6*], which requires the public authorities to give all relevant information to the public concerned 'as soon as it becomes available'." And in terms of the EIA Directive 2011/92/EU Article 6(4): "The public concerned shall be given early and effective opportunities to participate in the environmental decision-making procedures referred to in Article 2(2) and shall, for that purpose, be entitled to express comments and opinions when all options are open to the competent authority or authorities before the decision on the request for development consent is taken."

65. *RSPB v. Inch Cape* [2016] CSOH 103, July 18, 2016.

The "public concerned" includes "non-governmental organisations promoting environmental protection and meeting any requirements under national law" [EIA Directive 2011/92/EU, Articles 1(2)(e) and 11(3); Aarhus Convention Article 2(3)(a); *The Aarhus Convention: An Implementation Guide* (United Nations, New York, 2000), 107].

[5] Grid

As above.

[6] Other Concerns and Legal Regime

Flight Safety

The Civil Aviation Authority (CAA) ensures that the construction of wind turbines does not impact aviation in the UK. The CAA represents civil aviation interests on the DECC-chaired Aviation Management Board.

Habitat and Species Protection

The key instruments that are engaged by wind farm projects for the purposes of habitat and species protection are the *EU Habitats Directive 1992* and the *EU Birds Directive 2009*. As with all other EU laws, the application of these Directives will be under review as a result of the UK's plans to withdraw from the EU.

Marine Pollution

Offshore, a marine license may be required from the MMO. The MMO carries out licensing and enforcement functions on behalf of the Secretary of State in respect of all waters adjacent to England and all UK offshore waters, except those adjacent to Scotland. The equivalent organization under the jurisdiction of the Scottish Government is Marine Scotland.

§2.04 CHALLENGES AND SOLUTIONS

[A] Institutional Design: Challenges and Solutions

[1] Challenges

The purpose of the institutional framework for the application for development consent is designed to deliver decisions that take into account the various interests at stake.

In order to achieve an appropriately balanced outcome, the decision-making process is informed by a number of actors, as outlined in section §2.02 of this chapter, which brings a variety of different interests and concerns to the table.

This presents the decision-making authority (ultimately, the Secretary of State) with the challenge of objectively judging the information presented to it, whilst simultaneously deciding the weight of each of the interests and the evidence brought forward in support thereof.

Although the legal and institutional framework is designed to deliver a balanced, reasonable, and informed outcome, the need to make a decision inherently requires the decision-maker to make a judgment. Therefore, it is the role of the law to create a sufficient level of transparency, and by doing so providing insight into the way in which the final decision was reached. The challenge is to transform the subjective process of making a decision into objective elements which can be presented as reasons why the ultimate outcome of the decision-making process is balanced and reasonable.

[2] Solutions

The first step in identifying or creating objective elements of decision-making in the application process is to identify the way in which institutions interpreted and weighed the information presented to them in previous applications. From this, an analysis of the interpretative act of using information to reach certain conclusions can be created. If required, such an analysis could subsequently be used to set up guidelines for decision-makers to follow or take into consideration in future applications.

Some work in this direction has already been done by Maria Lee, who identified four categories of knowledge claims presented to decision-makers in the application process and the way in which they interact.[66] These categories are (i) technical or expert knowledge, (ii) prior institutional knowledge, (iii) local or lay knowledge and (iv) professional knowledge claims.[67] Following a number of case studies, it was discovered that decisions heavily rely on technical and professional knowledge claims while sidelining local or lay knowledge. As a result, actors presenting information that falls in the latter category have more chances of success if they manage to phrase their information as technical knowledge. Additionally, local or lay knowledge can achieve more weight in the decision-making process if presented to the decision-maker indirectly through prior institutional knowledge. This type of knowledge claim has significant weight in the decision-making process because the information presented has already been through an institutional process. As a result, local knowledge that was taken into consideration in a previous institutional process and is presented as part of the package of a prior institutional knowledge claim can have a more significant impact on the decision than if it were presented as a local or lay knowledge claim.

Further research into this topic could increase insight into the decision-making process, and if current practices prove to be satisfactory they could be developed into guidelines for decision-makers.

---

66. Lee (n. 2).
67. *Ibid.*

[B] Incentives: Challenges and Solutions

[1] Challenges

Creating a low-carbon energy market provides a significant financial challenge. An important element in attracting investment for offshore wind projects is the existence of long-term market certainty. This includes a level of certainty about the commercial viability of offshore wind projects and continued demand in the long run.[68] Additionally, it requires that the government sets out long-term policies that demonstrate the continued commitment required for the successful development of the offshore wind industry.

Balancing the need to provide clear long-term policy commitments to attract investment against the necessity of letting the market run its course in order to achieve the most cost-effective solution is a considerable challenge. Without long-term commitments investors lack the incentive and confidence to develop offshore wind. However, the government is reluctant to commit to long-term policies as it is concerned that interference with the free market will be detrimental to achieving the most cost-effective approach in its switch to a low-carbon energy market. Furthermore, the government is concerned about committing to more long-term policies now, when low-carbon technologies continue to be advanced and developed.

[2] Solutions

In order to achieve the required investor confidence the government stated that it would work to increase developers' understanding of the existing price support system under the EMR program and of the process for allocating Contracts for Difference.[69] This is helpful as it increases visibility of existing policies and financial incentives.[70] However, this does not address the need for clearer, more long-term commitments required to increase investor confidence.

Despite various financial incentives as outlined in §2.03[A][2] of this chapter, investor confidence remains relatively low. In order to unlock the full financial potential of offshore wind energy the government needs to create policy that sends a strong signal to investors that it is committed to offshore wind both in the near and not so near future. The Climate Change Committee has therefore advised the government to:

(i) extend funding under the Levy Control Framework to at least 2025;

68. Offshore Wind Industrial Strategy Business and Government Action, p. 14 https://www.gov.uk/government/uploads/system/uploads/attachment_data/file/243987/bis-13-1092-offshore-wind-industrial-strategy.pdf.
69. *See* for example the recently updated policy paper on Contracts for Difference published March 13, 2017, https://www.gov.uk/government/publications/contracts-for-difference/contract-for-difference.
70. As recommended by the Climate Change Committee in Offshore wind: A valuable ingredient in the UK's future energy mix December 2, 2015, http://tinyurl.com/h7fh3uq.

(ii) to reserve some money for offshore wind technology when it next sets out its timetable and funding pots for the next round of auctions;
(iii) to set auction reserve process for low-carbon options that have reached maturity;
(iv) to work with Ofgem and the National Grid to ensure flexibility options are able to capture their full value while low-carbon options face full integration costs; and
(v) to develop approaches to secure low-cost generation at the end of contract life and from repowering.[71]

[C] Regulations: Challenges and Solutions

With regard to regulations the UK faces two interesting challenges. First, there is the question of how the UK will adapt its regulations post Brexit. Leaving the EU could, for example, have an impact on the regulations with regard to application of the EIA regime. Currently the UK's plans with regard to adapting regulations post-Brexit are not known.

A second regulatory challenge the UK faces relates to the concept of landscape in the process of obtaining planning permission. When an application is being evaluated the impact of the project on the landscape is taken into consideration.[72] Although the concept is increasingly being given shape in various official statements, there is as of yet no clear definition suitable for straightforward application.[73]

In the specific context of offshore wind development, it should be noted that the relevant National Policy Statement[74] does not give any indication how to interpret or apply the concept of landscape in the assessment of visual impact of individual projects. This is especially notable when considering the significance that was given to visual impact in the decision to reject the application of the Navitus Bay project.

§2.05 CONCLUSION

The UK has already achieved considerable success in the development and deployment of offshore wind. The reasons are various, and certainly include the UK's enviable wind resources. However, and in addition to these a sound, responsive, and predictable regulatory regime has been arrived at which has created world leading investment certainty. However, with the advent of BREXIT there is the prospect of damaging uncertainty being introduced into what is currently a relatively benign regulatory environment. Beyond change to the EU law instruments discussed above, the UK's withdrawal from EU obligations entails the possibility of its release from the EU

71. Power Sector Scenarios for the Fifth Carbon Budget, https://www.theccc.org.uk/wp-content/uploads/2015/10/Power-sector-scenarios-for-the-fifth-carbon-budget.pdf.
72. *See* strategic environmental assessment (SEA) and Siting Related Planning III(2)(2).
73. Lee (n. 2) 7–9.
74. NPS EN-1.

Renewable Energy Directive, a change to its energy mix, the loss of access to existing EU-wide support schemes (but also the prospect of new ones as EU state aid rules may subject to change). The EU's complex merger control regime might be relaxed by a non-EU UK, and more pressingly, there is an open question as to whether EIB financed credit agreements may have their "material adverse change" clauses triggered by BREXIT. Will the UK be able to satisfy credit agreements containing provisions requiring compliance with EU law (i.e., EIB financing)? These are just a few examples of the changes that BREXIT may bring to what has been to date a highly advanced and successful model of offshore wind development.

PART II Offshore Wind Power of Developing Countries in Europe

CHAPTER 3

Legal Framework to Develop Offshore Wind Power in France

Catherine Banet

§3.01 INTRODUCTION: POLICY BACKGROUND

[A] France Energy Generation Profile

Offshore wind is still in its infancy in France, a country where nuclear energy dominates the energy generation mix. Meanwhile, new dynamics are progressively transforming the energy landscape, supported by international, European and national legal commitments to reduce greenhouse gases (GHG) emissions and to increase the share of renewable energy sources in the energy mix. France has now set ambitious goals and claimed industrial ambitions in the offshore wind sector.

In 2017, France still derives about 75% of its electricity generation from nuclear energy.[1] This historic choice represents a series of strategic advantages. First, in terms of climate policy, this makes France one of the industrialized countries that emit the least volume of GHG emissions per capita in the world (as far as concerns Metropolitan France).[2] Second, it aims to secure the electricity supply of the country at comparatively low prices. Meanwhile, the colossal investments required for upgrading the

1. With 78.2% of the electricity production in December 2015, the share of nuclear energy has never been so high since 2004. Conjoncture énergétique Décembre 2015, Commissariat Général au Développement Durable, no 728, February 2016.
2. It is important to note that the situation is very different in Corsica and overseas territories, where coal and low-power fuel oil still take an important part in electricity generation. In these parts of the French territories, renewable energies will make an active contribution to the reduction of greenhouse gases emissions. *See also* the national complement submitted by France and dedicated to the oversea territories in addition to the Nationally Determined Contribution of the European Union, under the Paris Agreement to the UNFCCC (submitted on October 5, 2016 as INDC).

nuclear reactors in the coming years (so-called "grand carénage") and the growing competiveness of electricity generation based on renewable energy sources - although not yet able to replace the volume of nuclear electricity production - is progressively changing the terms of the debate.[3] There is now a pledge to limit the share of nuclear energy in electricity generation to 50% by 2025, a target defined in the 2015 Energy Transition Law for Green Growth.[4]

Indeed, interesting changes in the energy mix are taking place, partly spurred by the French government's commitment made in 2008 to increase the share of renewables. The national targets defined by the EU legislation in the renewable energy sector have been working as a major driver. In Directive 2009/28/EC France committed to a target of a 23% share of renewable energy in its final energy consumption in 2020. This corresponded to an increase of 38% compared to the goal set by the 2001 RES-Directive.[5] Until 2008, France has been lagging behind in its follow-up of the EU targets. In 2004, renewable energy sources represented 12.5% of total electricity generation, but a closer picture shows that most part originated from hydropower (12%) and that only 0.5% came from "other renewables."[6] Four years later in 2008, at the time of the final negotiations of Directive 2009/28/EC, the situation had barely changed, with a 12.4% share of hydropower in the total electricity generation, 1% for wind and 0.7% for the other renewable energy sources, for a total of 14.1% for all renewables.[7] With the new commitments taken at EU level as backdrop, France started implementing the first support measures to increase its share of renewable energy sources in the energy mix. The adoption of the 2009 Grenelle Law I, the 2010 Grenelle II and the 2010 NOME Law on electricity defined a new regulatory framework to that respect. It is also in 2009 that the Grenelle of the Sea identifies the need for reviewing the legislation in order to facilitate the development of marine renewable energies.[8]

Today, France is still lying behind its neighbors to meet the 2020 targets under Directive 2009/28/EC. In 2015, renewable energy sources accounted for 14.9% of the final energy consumption. Although this already represents an increase by 48% compared to 2005, meeting the EU target and developing sectors such as offshore wind clearly requires a much more proactive government policy.

This slow start has to be put in perspective with more recent developments in France. Renewable energy sources are now regarded as a real option in the efforts to

3. Court of Audit, *Annual report*, 2016; Court of Audit, *Les coûts de la filière électronucléaire*, January 2012.
4. Energy Transition Law for Green Growth no 2015-992 of August 17, 2015, JOFR n°0189 of August 18, 2015, p. 14263. Target defined in Art. L. 100-4.-I.-5 of the Energy Code.
5. *See* National action plan for the promotion of renewable energies 2009–2020. In accordance with Art. 4 of European Union Directive 2009/28/EC).
6. Source: RTE. Reported in *Commission de Régulation de l'Energie, Rapport transmis à la DG TREN*, 2005. p. 48.
7. Data source: RTE. Reported in *Commission de Régulation de l'Energie, Rapport transmis à la DG TREN*, 2009. p. 92.
8. *Blue book on the commitments from the Grenelle of the Sea* (Livre Bleu des engagements du Grenelle de la Mer (July 2009), Ministry of Ecology, Energy, Sustainable Development and the Sea, p. 8.

both reduce GHG emissions and ensure security of energy supply by reinforcing local generation.

In terms of legal commitments, the 2015 Energy Transition Law for Green Growth defines a new target of 23% by 2020 and 32% in 2030 for the share of renewable energy sources in the final gross energy consumption. As concerns electricity generation, the target is a share of 40% of renewable energy by 2030.[9] Those targets are inserted into the Energy Code, which is a legal code gathering the various oil, gas and electricity provisions into one piece of legislation.

In June 2016, the French parliament passed a Blue Economy Law, with the ambition of developing economic activities at sea while preserving the marine environment.[10] Although this new law contains some provisions in favor of offshore wind installations (*see* Section §3.03[A][3]), the main piece of legislation defining objectives in the area of offshore wind development remains the energy legislation.

Looking at the latest figures, France is currently one of the EU's largest producers and consumers of renewable energy, with for instance the share of installed hydropower capacity close to 20% in 2014. Turning to consumption, renewable energy sources represented 14.6% of the final energy consumption in 2014 (18.4% for electricity, 18.1% for heating and 7.7% for transport)[11] and 14.9% in 2015.[12] Particularly during the last few years, the renewable energy sector has witnessed high increase rates particularly in the onshore wind sector and solar sector, which reflects the new dynamics at work. In the wind sector along, the installed capacity has increased by 29% in 2015 compared to 2014. In March 2017, the total generation capacity of wind power in France was 12.1 GW and represented 4.8% of the national electricity consumption.[13] However, those figures only relate to onshore wind and the progress made in offshore wind are still slow. At the beginning of 2017, there was still no offshore wind project in operation. The French government has recently decided to raise its level of ambition in the offshore wind sector, based on a combination of climate, energy and industrial policies. The first offshore wind projects are planned to come into operation between 2018 and 2020.

9. Article L. 100-4.-I.-4, Energy Code.
10. Blue Economy Law of June 20, 2016 (Loi n. 2016-816 pour l'économie bleue. JORF n.0143 du 21 juin 2016).
11. In 2014, France had 21,002 thousands toe or primary production of renewable energy from renewable sources. According to technology, the share was: 2.9% solar energy, 63.1% biomass and waste, 1% geothermal, 25.7% hydropower, 7.1% wind energy (Eurostat 2016).
12. Ministry for the Environment, Energy and the Sea, *Chiffres clés de l'énergie, Edition 2016* (Feb. 2017).
13. Ministry for the Ecology and Inclusive Transition, Scoreboard: wind power, first semester 2017, May 2017.

[B] Rationale for Moving from Onshore to Offshore: National Policy

[1] Assessing the Potential

The level of underdevelopment of the offshore wind sector in France contrasts with its potential and the role to be played by offshore wind in the future on a global scale.

With three coastlines of together 3,500 km and the second largest wind resources in Europe, France benefits from favorable geographical conditions for developing offshore wind.[14] The three coastlines of Metropolitan France include the Atlantic Ocean, the Channel Sea and Mediterranean Sea Basin. On those, offshore wind farms can be located in the territorial waters or in the exclusive economic zone (EEZ). According to the national trade union of renewable energies (*Syndicat des Energies Renouvelables*), France's offshore wind potential is estimated at 90 TWh. France Energie Eolienne (FEE), the industrial association for wind energy, estimates the potential to be 80 GW for bottom-fixed offshore wind for an area of 10,000 km^2 and 140 GW for floating offshore wind for an area of 25,000 km^2.

The geography and topography of the French coasts may represent a challenge based on today's technologies. The continental shelf is relatively narrow and rapidly becomes deep water. Not surprisingly, most wind parks under development are located in the territorial waters. For the same reasons, France sees a huge potential in floating offshore wind turbines, the testing of which is strongly supported by the government.

[2] Multiple Motivations

The first motivation for developing offshore wind in France is the contribution it will make to the increase of renewable energy sources in the energy mix as well as target compliance both at EU and national levels.

Second, France intends to develop strong national industrial competences within the offshore wind sector, including the construction and service industry. New factories have been opened during the past few years, like in 2015 in Montoir-de-Bretagne, Saint-Nazaire, or are planned, like the blade factory for offshore wind turbines in Cherbourg which started construction in March 2017.[15]

Third, the development of onshore wind projects in France is opposed by local population. Going offshore is seen as a solution to overcome that opposition. Meanwhile, even offshore wind projects have been facing opposition when located too close to shore, with numerous appeals to permit and legal proceedings delaying the start of the projects. A solution is to move further offshore. It is in that context that floating wind projects have been identified as a priority area.

A final motivation is to boost the development and production of renewable energy in French overseas territories, which remain highly dependent on fossil fuels for some of them.

---

14. If the coast of the oversea territories is added, France has the second largest maritime area in the world after the United States, with 11 million km^2.
15. Factory plant of the Danish manufacturer LM Wind Power.

[C] Policy Setting: The Role of Wind and Offshore Wind Power in Policy Papers

[1] Latest Policy Developments

A policy on marine renewable energy (MRE) resources, including offshore wind, started to develop along the 2009 Grenelle of the Sea (*Grenelle de la Mer*). The Grenelle of the Sea is copied on the model of the other Grenelle processes in France (originally a model developed for social policy) and works on the basis of a wide debate with multiple stakeholders with the purpose of gathering opinions.

The Grenelle of the Sea started discussing targets for offshore wind parks (around 15 GW bottom-fixed installations by 2030) and launched a series of initiatives, many of them oriented towards the review of marine planning procedures and technology development. The research and development (R&D) was built on the experience from the National Technology Platform for Marine Renewable Energies (IPANEMA),[16] and was supported by research programs managed by the environment agency ADEME. Another follow-up initiative was the launch by the government in November 2013 of a new roadmap for both fixed-foundation offshore wind farms and floating wind power installations. The review of marine planning procedures is addressed in Section §3.03[B][4] below.

The current targets for offshore wind are defined in two main documents completing the general targets of the 2015 Law on Energy Transition for Green Growth: the Multiannual Programme for Energy (PPE) and the National strategy for the sea and the coastal area (SNML). Those two strategy documents are analyzed below. Although the objectives remain common, the government defines separate strategies, permitting rules and support measures for the different sources of renewable energy at sea, namely: bottom-fixed offshore wind; floating offshore wind; and MRE sources (wave power, tidal power, osmotic power, ocean thermal power).[17]

[2] Quantitative Target Setting: Multiannual Programme for Energy (PPE)

The first main document defining quantitative targets for offshore wind energy policy is the Multiannual Programme for Energy (*Programmation Pluriannuelle de l'Energie*, PPE), adopted by Decree on October 28, 2016.[18] This PPE covers Metropolitan France, the oversea territories being subject to separate PPEs.

The adoption of the PPE is a requirement introduced by the 2015 Energy Transition Law for Green Growth,[19] now codified in the Energy Code.[20] The PPE

16. A cluster to support innovation in marine renewable energy technologies created in 2008.
17. In principle, offshore wind is not a source of a marine renewable energy since it is not based on the use of the water but of the use of wind. Meanwhile, many national policy documents include offshore wind power within the category of marine renewable energy sources.
18. Décret n. 2016-1442 du 27 octobre 2016 relatif à la programmation pluriannuelle de l'énergie.
19. Article 176, Law no. 2015-992 of August 17, 2015 on energy transition for green growth.

supplements the Multiannual Programme for Investments (PPI) of April 24, 2016. Indeed, ahead of the adoption of the new PPE and to avoid investment uncertainty, the Energy Ministry adopted on April 24, 2016 an Administrative Order detailing the objectives both in general terms and by renewable energy type. This included a general commitment to increase by 50% the installed capacity for renewable energy sources for electricity generation. Those targets were confirmed in the PPE.

The PPE defines concrete implementation measures and quantitative targets designed to reach the objective defined in Articles L. 100-1, L. 100-2 and L. 100-4 of the Energy Code. The PPE defines the roadmap for the State for two successive periods: 2016–2018 and 2019–2023.[21] The development of marine renewable energies (including offshore wind)[22] is a priority of the PPE.

For offshore wind, the PPE defines the following objectives:

<table>
<tr><th>Deadline</th><th>Total Installed Capacity</th><th>Awarded Projects</th></tr>
<tr><td colspan="3">Bottom-fixed installations</td></tr>
<tr><td>December 31, 2018</td><td>500 MW in operation</td><td rowspan="2">Between 500 and 6,000 MW of additional capacity, depending on the fulfillment of the following criteria: consultation on the identification of the most suitable areas, feedback on implementation of the first projects and price conditions.</td></tr>
<tr><td>December 31, 2023</td><td>3,000 MW in operation</td></tr>
<tr><td colspan="3">Floating wind mills and marine current power</td></tr>
<tr><td>December 31, 2023</td><td>100 MW</td><td>Between 200 and 2,000 MW of additional capacity, depending on the fulfillment of the following criteria: feedback on implementation of the pilot wind farm projects and price conditions.</td></tr>
</table>

The PPE defines the calendar for the launch of call for tenders for each technology type.

20. Article L. 141-3, Energy Code.
21. The PPE is of a five-year duration and should coincide with the duration of the presidency of the French Republic. Therefore, the first PPE only covers three years, and then a standard period of five years.
22. *See* comment in footnote 21.

[3] *National Strategy for the Sea and the Coastal Area* (Stratégie nationale pour la mer et le littoral, *SNML*)

The second main guidance document for marine renewable energies, including for offshore wind, is the national strategy for the sea and the coastal area (*Stratégie nationale pour la mer et le littoral*, SNML). The national strategy was released on February 24, 2017 and was formally adopted by Decree.[23] It defines the framework of action for public authorities on sea and coastal issues. It defines primarily qualitative targets for offshore wind development, building on national ambitions and a list of actions to be performed.

The Strategy suggests a series of indicators for assessing progress in achieving structural objectives such as: ecological transition of the sea and the coastal zone; development of a maritime economy; preservation of the ecologic status of the marine environment; and the international influence of France at the international arena. The Strategy also defines a list of priority actions. As far as offshore wind is concerned, the following priority actions are of particular relevance:

- Support innovation in the maritime sector, increasing research capacity (nr 2).
- Elaborate maritime spatial planning to reconcile uses, seek synergies between activities and integrate new activities (nr 6).
- Establish 100 Positive Energy Marine Territories (nr 7).[24]
- Meet France's ambitions for renewable marine energies (nr 12).
- Support innovative pathways and energy transition in maritime transport and services (nr 16).

[4] *Follow-Up Reports on Implementation Strategy*

On March 21, 2017, the French MP Catherine Troallic submitted to the Minister in charge of the environment, energy and the sea, on the request of the Minister herself, a report on energy transition of the industrial maritime territories.[25] The report looks at the question of the consequences on the energy transition on the territories which are located on the coast and which are economically heavily dependent on maritime industries and related services. The report defines twelve concrete proposals to enable a "fair transition."

On October 26, 2016, an information report on the implementation of the 2015 Energy Transition Law for Green Growth was published by the French National Assembly.[26] The report reviews the progress made in the implementation of the new law, but also comes with a critical assessment of the effects of the new law on the

23. Decree No. 2017-222 of February 23, 2017 on a national strategy for the sea and the coastal area.
24. Positive Energy Marine Territories (*territoires maritimes à énergie positive*).
25. Report, C. Troallic, "Territoires industriels maritimes: 12 propositions pour réussir la transition juste", March 2017.
26. Information Report no. 4157 on the application of Law no. 2015-992 of August 17, 2015 on Energy Transition for Green Growth, presented by J.-P. Chanteguet.

energy transition process. As concerns offshore wind energy, the report takes note of the process made in terms of marine planning and permitting procedures, but also points out the still slow development of the sector and the remaining conflict areas (e.g., by the limitation put by military activities on access to suitable areas; *see* Section §3.03[B][8] below).

§3.02 INSTITUTIONAL DESIGN FOR OFFSHORE WIND

[A] The "Main" Authority During the Development of Offshore Wind Power

Ministry for the Ecological and Inclusive Transition (*Ministère de la Transition Ecologique et Solidaire*)[27] covers a large portfolio covering: sustainable development, environmental protection, green technologies, energy transition, energy policy in general including tariffs, climate, natural and technological risks prevention, industrial safety, transport, the sea. It drafts and applies the policies for those different areas of competence.

The Commission for Energy Regulation *(Commission de Régulation de l'Energie, CRE)* is the French regulatory authority for gas and electricity. Established in 2000, its role is to ensure the proper operation of the electricity and gas markets, in accordance with the energy policy's objectives and for the benefit of final consumers. CRE has a large range of competences, including: ensuring access to the public electricity grid; ensuring the proper operation and development of the electricity grid and infrastructure; ensuring the independence of network operators; contributing to the development of the European electricity and gas markets; and contributing to the implementation of electricity supply and generation policies. It monitors electricity, gas and CO_2 markets. It administers some parts of the support schemes for renewable energies. CRE administers the public service contributions, which cover the financial support to renewable energy source in the framework of the support schemes. As part of the offshore wind tender procedure, the list of award criteria is first submitted to CRE for advice, and the tendering procedure is organized by CRE.

[B] The "Supplementary" Authorities During the Development of Offshore Wind Power

[1] Central Government Level

[a] Energy and Environmental Authorities

The Environment and Energy Management Agency (*Agence de l'Environnement et de la Maîtrise de l'Energie*, ADEME) is a public agency. It is jointly supervised by the Ministry for the Ecological and Inclusive Transition and the Ministry of Education and Research.

27. The Ministry has changed names almost each time a new minister has been appointed.

Its role focuses on energy efficiency, renewable energy and sustainable development in broad terms. ADEME has regional offices.

Préfet - The Préfet is the representative of the state in the department. He/she has some competences in terms of permitting. In offshore wind projects, the Préfet of the department and the Préfet of the maritime area (Maritime Préfet) are the responsible authorities for coordinating the consultation procedures in the process of identifying the suitable areas for offshore wind energy projects.

[b] Marine Authorities

In addition to the leading role of the Ministry for the Ecological and Inclusive Transition, coastal and maritime policies are coordinated by the National Council of the Sea (*Comité Interministérial de la mer* - CIMer) and by the General Secretariat for the Sea (*Secrétariat Général de la Mer* - SGMer). Both bodies are under the direct responsibility of the Prime Minister.

[2] The Role of the Local Government

Although the renewables target binds the State at the national level, the implementation of the objectives is delegated to the local level, with a strong involvement of local authorities.

When mapping the suitable areas for offshore wind projects, the consultation process includes local municipalities. Coastal municipalities are also entitled to financial compensation for the visual interference from the offshore wind mills.

[3] Other Players Vital for the Development of Offshore Wind

[a] Industry Associations

The main industrial associations representing the interest of the offshore wind energy sector are:

- France Energie Eolienne (FEE)
- France Energies Marines (FEM)
- Observatoire des énergies de la mer
- Syndicat des Energies Renouvelables (SER), which is an association for renewable energy in general.

[b] Commercial Actors

Among the key commercial actors present in the offshore wind sector in France are both national and international companies. Among the French national companies are: EDF Energies Nouvelles (partially state owned), Engie, Alstom, Areva (despite recent change in priorities). Certain companies with both national and foreign mother companies have been created in the sector, such as Éolien Maritime France (EMF), owned notably by EDF and Dong Energy Power, or Ailes Marines SAS owned by Iberdrola and EOLE-RES SA. Foreign companies have also established subsidiaries in France, such as WPD offshore France, a daughter company of the German group WPD Offshore.

[c] Grid Operators

Created in 2000, *Gestionnaire du Réseau de Transport d'Electricité (RTE)* is the transmission grid operator (TSO) and a subsidiary of EDF. It has received a mission of public service which is to ensure fair access to the electricity network for all participants in the electricity market. RTE operates, develops and maintains the French high-voltage electricity transmission network. It must preserve the operational security of the electricity system and maintain power balance. Offshore wind energy projects will fall under the area of competence of RTE. RTE is consulted ahead of the opening of competitive tendering procedures for offshore wind energy projects and is often involved in the negotiations in the framework of the competitive dialogue-procedures. Several agreements will need to be entered into between the offshore wind energy project developer and RTE, including a connection agreement.

Enedis (formerly Electricité Réseau Distribution France, ERDF) is the main distribution system operator (DSO) in France, at 95%. The remaining 5% is managed by local distribution companies. Enedis is a wholly owned subsidiary of EDF. It is not expected that DSOs will be directly involved in the offshore wind projects currently tendered, since the projects will be primarily connected to the high voltage grid.

[C] The Role of "Law" under the Institution Decision Making and Related Policies

The French legislation applicable to offshore wind projects has evolved considerably during the last few years, with the purpose of enabling those projects.

§3.03 LEGAL DESIGN FOR OFFSHORE WIND

[A] Incentives

[1] Main Finance Scheme

Offshore wind in France is a technology which still needs public support in order to develop. The support policy differs between offshore wind with bottom-fixed installations (a) and floating installations (b).

The evolution of the financial support scheme in favor of renewable energy is highly influenced by EU law and EU state aid law. While Article 107 of the Treaty on the Functioning of the European Union (TFEU) defines the general principles of state aid policy, the Guidelines on State aid for environmental protection and energy (EEAG) (2014–2020) provide for the key principles applied by the European Commission when reviewing the compatibility of national support measures in favor of renewable energy.

[a] Bottom-Fixed Offshore Wind

In principle, access to support for electricity generation based on renewable energy sources is subject to two different procedures: competitive tendering procedure, such as the call for tenders and competitive dialogue; and a regime of "opened counters" (*guichets ouverts*), which will deal with the demand of qualified RES-operators.[28] In principle again, both procedures give right to access to two different remuneration mechanisms: the power purchase obligation by EDF (*obligation d'achat*)[29] or the newly established remuneration complement contract (*complément de rémunération*).[30] Although based on the same general principles, the support regime for offshore wind energy in France differs in practice.

When the market is not delivering the additional generation capacity needed to meet the national targets defined in the PPE, the Minister in charge of energy has the competence to launch calls for tenders.[31] This has become de facto the standard procedure for developing offshore wind in France. This does not prevent offshore wind projects to develop outside the procedure for call for tenders. Meanwhile, this has not been the case so far, the absence of project having motivated the government to launch specific tenders. Therefore, the development of offshore wind projects in France rely in practice on calls for tenders, although it can also take place outside those procedures in theory.

[i] Tender Procedures and Tendering Rounds

The Ministry has launched *3 tendering rounds* for the building of *fixed-foundation offshore wind farms*.

Round 1 – The first call of tenders was launched in July 2011, and the tenders were awarded in 2012 to four projects (four concessions) for a total target capacity of 2,000 MW:

(1) Fécamp (Seine-Maritime, 498 MW), awarded to Eolien Maritime France.
(2) Courseulles-sur-Mer (Calvados, 450 MW), awarded to Eolien Maritime France.
(3) Saint-Nazaire (Loire Atlantique, 480 MW), awarded to Eolien Maritime France.

28. The projects which are not subject to tendering rules and which want to benefit from the available support schemes are subject to the regime of "opened counters" (*guichets ouverts*), which will deal with their demand.
29. Article L. 314-1 to L. 314-13, Energy Code.
30. Article L. 314-18 to L. 314-27, Energy Code.
31. Article L. 311-10, Energy Code.

(4) Saint-Brieuc (Côtes d'Armor, 496 MW), awarded to Ailes Marines SAS.

An additional fifth project was initially shortlisted, but the project was abandoned by the authorities because the price proposed by the tenderers was too high (Le Tréport, Seine-Maritime – Somme). This project is now part of the second tendering round.

The awarded offshore wind parks should have been completed by 2018, but the entry into operation is now planned for 2020–2021. Eolien Maritime France is a consortium between EDF and engineering company Alstom and won three of the four sites, at Fécamp, Courseulles-sur-Mer (Normandy) and Saint-Nazaire (Loire). Ailes Marines SAS is a consortium between Iberdrola and the nuclear conglomerate Areva.

Round 2 – The second round of tendering procedures was launched in March 2013 in two different areas: off the town of Le Tréport in northern Normandy and off the islands of Noirmoutier and Yeu on the Vendee coast. The tender has a total capacity of 1,000 MW (500 + 500) and was awarded in May 2014 to a consortium group managed by French gas and energy company Engie (two concessions). The Engie consortium comprises Portuguese EDP Renovaveis, French Neoen Marine and Areva. The winners of the tenders plan to apply for the necessary permits in the Spring 2017. The two parks should enter into function between 2020 and 2023 and will comprise around 200 windmills.

Round 3 – The third tendering round was launched in April 2016. The turbines will be construed on one sole area, which is located along the coast near Dunkirk. The offshore wind park to be developed must have a capacity between 250 MW and 750 MW. It is a requirement that the project contributes to local economic growth. A particularity of the third round is that it applies a new procedure of "competitive dialogue" (*dialogue concurrentiel*) (*see* details below). The government has also started a simplification work for the call for tenders. Tender awards are based on three criteria: price, project development and environmental impact, in that priority order.

Round 4 of the tender procedures is under preparation. The identified area is close to Oléron and is of approximately 120 km^2. In according with the methodology put in place during Round 3 of the tender procedure, a series of preliminary studies will be performed in 2017 before the tender procedure is launched. Even if the procedure of the former round has started relatively recently, the Ministry deemed it important to announce this new tendering round in order to keep a regular pace in the development of offshore wind in France.

[ii] Evolution of the Competitive Tendering Procedure Over Time: From Traditional Call for Tenders to the Competitive Dialogue

The tendering procedure has evolved since the first tendering round was launched. Between the second and third tendering rounds, at the occasion of the first meeting of the National Committee for renewable energies at sea (*Comité national des énergies renouvelables en mer*, CNEM), the French authorities started a process for the elaboration of a roadmap for offshore wind bottom-fixed installations. This process has resulted in changes in the methodology applied to the tendering procedure with the purpose of reducing delays, increasing competition, reducing project costs, limiting conflicts in area use and ensuring local acceptance. The new methodology has been applied since the third tendering round.

The first change in the methodology is the introduction of a better mapping of the relevant geographic areas and a better planning of the consultation work ahead of the launch of the tender. The center for studies and expertise on risks, environment, mobility and city and land planning (*Centre d'études et d'expertise sur les risques, l'environnement, la mobilité et l'aménagement*, CEREMA) conducted a study which identified the most relevant areas for the development of offshore wind (technical study). The French TSO, RTE, performed another study which identified the conditions for the connection of the offshore installations to the electricity grid. Those two studies by CEREMA and RTE serve as basis for the launch of local consultation procedures. Those local consultations are organized by coastline and are coordinated by the Maritime Préfet and result in the identification of the most suitable zones for the development of offshore wind projects.

The second main change has been the introduction of a new procedure for the selection of the projects, the so-called "competitive dialogue"-procedure (*dialogue concurrentiel*).[32] This procedure replaces the traditional call for tenders, but is still a competitive tendering procedure. The purpose of the new procedure is specifically to reduce project costs and secure the realization of the project. The main stages of the competitive dialogue are:

(1) Phase 1:
   - The pre-selection of the tenderers on the basis of their technical and financial capacities.

(2) Phase 2:
   - Invitation sent to the selected candidate to start the dialogue-procedure on the basis of draft terms of reference with pre-selected tenderers;
   - the sending to the candidates of the final terms of reference at the end of the dialogue;
   - the submission of tenders by the tenderers.

(3) Final: Evaluation of the tenders and designation of the winner.

As part of the competitive tendering procedure, the list of award criteria is first submitted to the CRE for advice. CRE then coordinates the tendering procedure.

The European Commission's Guidelines on State aid for environmental protection and energy (2014–2020) (EEAG) contain new rules on competitive and strategic bidding, which also apply to wind energy. The EEAG do not explicitly distinguish between offshore and onshore wind. Pursuant to the EEAG, as of January 1, 2017, aid must be granted in a competitive bidding process on the basis of clear, transparent and

32. Article R. 311-25-1 to R. 311-25-13, Energy Code.

non-discriminatory criteria. This rule applies to wind energy installations of an installed capacity of more than 6 MW or 6 mills.

Access to the support schemes is either subject to a procedure of calls for tenders or of "open counters" system. Under both procedures, projects developers can benefit from two support schemes: the purchase obligation by EDF or the remuneration complement contract.

[iii] Financial Support Schemes: Power Purchase Obligation or Remuneration Complement Contract

In most cases, offshore wind projects are subject to tendering procedures. Pursuant to Article L. 311-12 of the Energy Code, the text of the competitive tendering procedure will precise the type of support scheme the tender will be based on, choosing between: a purchase obligation for the electricity generation based on a fix purchase tariff (alternative one) and a remuneration complement (alternative two). The choice is made by the administrative authority responsible for launching the tender (the Ministry).

Outside tendering procedures, offshore wind project developer can claim access to the purchase obligation or the remuneration complement mechanism, according to the conditions defined by Decree.

Alternative One – From Feed-In Tariffs to Feed-In Premium: Remuneration Complement Contract

For the offshore wind projects awarded in 2011 and 2013, the purchase price was not regulated as it was for onshore wind, but determined as part of the tendering procedure, based on the proposal put forward by the bidder. The purchase price was one of the assessment criteria for the bid and was fixed for the whole period (feed-in tariff). Again, no project developed outside the call for tender procedures.

After the entry into force of the EEAG (2014–2020), the situation changed. The general principle defined in the EEAG, and applicable as of January 1, 2016, is the one of a premium on the top of the market price, instead of feed-in tariffs system. From that date, aid must be granted as a premium in addition to the market price (premium) whereby the generators sell their electricity directly in the market (paragraph 124(a), EEAG). For wind energy, this rule applies to installations with an installed capacity of at least 3 MW or 3 generation units (paragraph 125, EEAG). Below this threshold, installations can still benefit from an FIT, but this is unlikely to apply to offshore wind projects in practice.

As a consequence of the entry into force of the EEAG, France has amended its financial support scheme for renewable energy sources. The new and main financial support regime is based on a remuneration complement contract (*complément de rémuneration*) which is a feed-in premium.[33] The remuneration complement is a bonus paid to renewable electricity producer in addition to the price he gets from the sale of

33. Article L. 314-18 to 314-27 of the Energy Code; Decree No. 2016-682 of May 27, 2016.

its electricity on the market. It applies to projects awarded after January 1, 2016 without retroactive effect.

The conditions set by the EEAG have been applied in the third tendering round for offshore wind in France which is based on the remuneration complement mechanism. EDF is in charge of entering into a contractual agreement (power purchase agreement) with the winner of the tender for the payment of the remuneration complement.[34]

Alternative Two – Power Purchase Obligation

For certain emerging technologies, France still maintains a power purchased obligation by EDF or local DSOs at a fixed price determined by Decree.

Projects which aim to benefit from the purchase obligation but which are not part of a competitive tendering procedure must obtain a CODOA-certificate from the Préfet, providing them the right to benefit from the purchase obligation (*certificate ouvrant droit à l'obligation d'achat*, CODOA).[35]

[b] Floating Offshore Wind

In its efforts to support the development of floating offshore wind projects along the French coasts, and in addition to R&D programs, the government is promoting demonstration and testing projects in pre-commercialization phase. For those, the government's support relies on a system of call for projects within a special investment program called "programme des investissements d'avenir." The winning projects are eligible to the power purchase obligation,[36] but not the remuneration contract[37] since the two mechanisms are not cumulative.

As of 2017, four projects of a total capacity of 24 MW each have already been awarded:

- "Provence Grand Large" is a project composed of 3 wind mills of 8 MW each. The project is developed by EDF EN, using turbines manufactured by Siemens and a floating structure by SBM/IFPEN. The project is located at Faraman, in the Mediterranean Sea.
- The "éoliennes flottantes Golfe du Lion" is a project composed of 4 mills of 6 MW. The project is developed by Engie/EDPR/CDC, using turbines manufactured by General Electric and a floating structure by Eiffage/PPI. The project is located at Leucate, in the Mediterranean Sea.
- The "éoliennes flottantes de Groix" is a project composed of 4 mills of 6 MW. The project is developed by Eolfi/CGN, using turbines manufactured by General Electric and a floating structure developed by DNCS in collaboration with VINCI. The project is located at Groix, in Brittany.

34. The contract is subject to French law.
35. Articles R. 314-6 et seq., Energy Code.
36. Article D. 314-15(7), Energy Code.
37. Article D. 314-23, Energy Code.

- "Eolmed" is a project composed of 4 mills of 6 MW. The project is developed by Quadran, using turbines manufactured by Senvion and a floating structure by Bouygues Travaux Publics and Ideol. The project is located at Gruissan, in the Mediterranean Sea.

These four demonstration projects are financially supported at a level of EUR 330 million each by the *Programme des investissements d'avenir* and in addition from the specific purchase tariff.

The government aims to launch tendering procedures for commercial floating wind energy projects in the near future. For that purpose, mandate was given to the Préfets coordinating the maritime façade to consult local stakeholders in order to identify the most appropriate areas.

[2] Supplementary Finance Scheme

[a] Low Interest Loan and Loan Guarantee

Not applicable.

[b] Tax Credit

Electricity generated from renewable energy sources is promoted through several tax incentives. Persons investing in renewable energy plants are eligible for an income tax credit (Crédit d'Impôt). The Finance Act 2005 created the Tax Credit for promoting sustainable development and energy savings. Meanwhile, since the measure mainly targets individual use, it is deemed not relevant for offshore wind energy projects so far.

Companies can also benefit from a research tax credit on their environmental investments, if the undergone activities correspond to eligible research activities. Research tax credit can be relevant for offshore wind energy projects.

[c] Favorable Grid Cost-Sharing Rules

For offshore wind projects which are subject to tendering procedures, the grid connection costs are in principle to be borne by the tenderer. However, the Ministry has indicated that, as part of the newly established competitive dialogue-procedure, the matter could be subject to negotiation.

[d] The Links to ETS, CDM, and JI

Not applicable.

[e] *Other Funding Source, Such as National Research Fund and Supportive Scheme*

ADEME provides advice, background information, and financial support to renewable energy projects, at different stages of research, development, and demonstration of technologies. ADEME has launched already numerous calls for proposals for R&D projects under the national program for future perspectives (Investissements d'Avenir), aimed to stimulate innovation. The projects benefit usually from a direct grant.

[f] *International Trade Law Concerns of the Subsidy*

The terms of the tenders launched by French authorities might raise issues under international trade rules, and in particular the World Trade Organisation regime. It is reported that the tenders require a certain percentage of local content requirements which will hardly find justification under the exceptions rules of the WTO Agreements.

[3] *Incentive for Construction Harbor, Construction Vessel and Grid, or Turbine Manufacturers*

[a] *Preferential Insurance Regime*

The law on the Blue Economy of June 20, 2016 introduces a new insurance regime for MRE Installations, with the aim of reducing insurance costs. Article 84 of the law now classifies MRE installations as being at "large risks," which excludes them from the mandatory regime applicable to terrorist risks and natural catastrophes. The consequence of this change is to let the market fix the terms of cover for MRE installations. This will ultimately reduce their insurance cost for MRE projects. Being qualified as "large risks," the MRE installations can choose the law applicable to their insurance policy. There is no longer an obligation to make them subject to French law.

[B] Regulations

Since the Grenelle Law II,[38] France has initiated several reviews of its legislation in order to streamline the licensing procedures applicable to electricity generation from renewable energy sources, including from offshore wind.

The provisions applicable to the development of an offshore wind energy project are to be found primarily in the Energy Code, the Environment Code, General Code on Public Property (CG3P) and, to a certain extent, the Town Planning Code.

In addition, project developers will be able to conclude purchase agreements for the electricity generated.

38. Law No. 2010-788 of July 12, 2010.

[1] *License Scheme*

Even if offshore wind projects have been awarded through a tendering procedure, the project is subject to the same requirements and must obtain the same permits that if the project would have been developed outside the tendering procedure.

[a] *Concession to Occupy Maritime Space and Seabed Within the Maritime Public Domain or the EEZ*

The legislation provides for two different legal bases for granting a concession to establish in French jurisdiction installations such as offshore wind mills and related infrastructures, including cables, dependent on their location, i.e., would it be in territorial waters or in the EEZ. While the first offshore wind projects are to be developed in territorial waters (within public maritime domain),[39] the French legislation enables their development within the EEZ too (beyond public maritime domain).

For a long time, France has been lacking implementation legislation for the development of, *inter alia*, renewable energy sources for electricity generation within the EEZ and on the continental shelf. Decree 2013-611 of July 10, 2013 clarified the licensing regime applicable to those circumstances.[40] It was adopted pursuant to Law 76-655 of July 16, 1976 relating to the Economic Zone off the Coasts of the Territory of the Republic, and which implements in French law the provisions of the UN Law of the Sea Convention. The Maritime Préfet of the concerned marine region is the competent authority for awarding the concession.

Within the public maritime domain, offshore wind mills and related transport installations such as cables must get a concession to occupy the public maritime domain, so-called maritime domain concession (*titre domanial d'occupation*) (Articles L. 2124-1 to 3 and R. 2124-1, CG3P). The project developer must pay a fee to fiscal authorities in relation to the public domain concession (*redevances d'occupation du domaine public*).[41] Additional procedural provisions have been adopted in Decree N. 2004-308 of 29 mars 2004 on concessions for use of the public maritime domain outside harbors. The Decree provides for the constitution of financial guarantees for decommissioning phase and restauration of the site. The Préfet is the competent authority for awarding the concession in the form of a prefectoral order. The draft prefectoral order to subject to the prior assent of the responsible Maritime Préfet as well as a wide hearing process with stakeholders and local authorities. It is also subject to a public inquiry. The concession, originally given for a duration of thirty years is now given for a duration of forty years.

39. It includes internal waters and the territorial sea. Baselines are calculated in accordance with Decree No. 2015-958 of July 31, 2015.
40. Decree No. 2013-611 of July 10, 2013 (*relatif à la réglementation applicable aux îles artificielles, aux installations, aux ouvrages et à leurs installations connexes sur le plateau continental et dans la zone économique et la zone de protection écologique ainsi qu'au tracé des câbles et pipelines sous-marins*).
41. Article L. 2124-1 to 3, CG3P.

For projects located both within public maritime domain and the EEZ, the French legislator has adopted the approach of a single authorization procedure (*autorization unique*), which includes both the generation units and the related installations. Those rules are defined:

- for projects within the public maritime domain, by Law N. 2014-1545 of December 20, 2014 (Article 18);
- for projects developed on the continental shelf or within the EEZ, by Law N. 2016-1087 of August 8, 2016 on the Recovery of Biodiversity, Nature and Landscapes.

The project subject to the first tendering rounds are all located within the public maritime domain.

[b] Authorization Pursuant to Water Resources Protection (Water Resource Protection License)

Pursuant to the Environment Code (Articles L. 214-1 and R. 214-1), installations at sea of a value of more than EUR 1.9 million must obtain a water resource protection license (authorization procedure). The application is usually made in parallel to the concession one by the project developer.

Although the authorization procedures seem to be limited to those two concessions, further authorizations have been needed in practice for the offshore wind projects awarded in the first rounds. For example, a building permit has sometimes been requested under the Town Planning Code. This indicates that the licensing regime needs further clarification, but also coordination between the different central and local public authorities involved in the licensing process. Decree 2012-40 provides now that renewable energy projects subject to maritime domain concession are exempted from city planning permit for all onshore related infrastructures and works (Articles L. 421-5 and R. 421-8-1, Town Planning Code).

Several simplification and coordination rules have been recently enacted.

First, environmental permits would be required in addition to the water resource protection license, notably due to the specificities of the site (fauna, flora), a procedure for single environmental permit has been tested since 2014 (*autorisation environnementale unique*). Such single environmental permit can be delivered by the Préfet. It aims to shorten the review period for the application and to deal more efficiency and rapidly with possible general appeals against the granted permit. The procedure has been extended as a general rule by the 2015 Energy Transition Law for Green Growth. Offshore wind energy projects will benefit from this single environmental permit.

A second simplification initiative has resulted in the adoption of Decree N. 2016-9,[42] which provides for new specific procedures for dealing with appeals against concessions given to offshore wind energy projects and related installations. The

42. Décret n° 2016-9 du 8 janvier 2016 concernant les ouvrages de production et de transport d'énergie renouvelable en mer.

Administrative Court of Appeal of Nantes is designated as the first and last instance to decide on all appeals and legal proceedings against the concession permits, the single environmental permit and the operating permit for the generation installations and related facilities and infrastructures. The competence of the Court extends to all authorizations and permits relevant for the development of an offshore wind energy projects. The same Decree extends the duration of the concession to occupy the public maritime domain (maritime domain concession) from thirty to forty years. Finally, the deadline for appealing a water resource protection license is limited to four months.

[c] Operating Permit for the Electric Installation

In general, all electricity generation installation based on offshore wind energy must obtain a construction and operating permit for electricity generation plan pursuant to the Energy Code (threshold of 50 MW, Article R. 311-2 and 5 of the Energy Code). However, when the project is awarded under a competitive tendering procedure, the winning bidder is exempted from applying to such permit and is automatically granted an operating permit (Article L. 311-11, Energy Code) and access to the financial support chosen for the specific tender. Operating permits for electricity generation installations are granted for twenty years (last amendment enacted by Decree 2016-9 of January 8, 2016).

All permits, concessions, or contracts are subject to French law. During the last tendering procedure subject to competitive dialogue, the place, governing jurisdiction and language of arbitration have been questioned. The public authorities have answered that the working language of the arbitration must be French, the arbitration proceedings must take place in a Member State of the European Union at the date of the demand for arbitration. Procedural rules can be agreed by parties or follow the rules in force at the place of arbitration.

[i] Decommissioning

The project developers must also ensure that the turbines will be removed after decommissioning and so commits to a general restoration requirement at the end of the occupation of the public domain. The project developer must inform the responsible licensing authority (normally the Préfet) of its intention to decommission the installation at least five years before the end of the operation phase. The conditions for decommissioning are defined in the concession. Pursuant to the Environmental Code, the operator needs to set aside a financial guarantee for that purpose (Article L. 214-2, Environmental Code). Bidders must therefore provide a financial guarantee, which vary according to the size of the project and the specificities of the site. For example, in the 2011 tender, it has been fixed to at least EUR 50,000 per turbine.[43]

43. *See* Specifications of tender no. 2011/S 126-208873.

[2] Environmental Impact Assessment (EIA)

Both the maritime domain concession and the water resource protection license require the performance of an EIA. The award of the concessions depends on the positive result of the EIA (Environmental Code, R. 122-1 and -2). Coordinating supplementary EIAs related to the same projects is part of the streamlining foreseen by the single environmental permit. The impact assessment performed for the project under one piece of legislation may serve the purpose of the EIA under the other legislation, provided that the two assessments contain the same information (Article R. 214-6, Environmental Code). When reaching its decision, the Préfet of the Department must get the preliminary consultative advice of the Environment Agency and integrate the comments given by the Maritime Préfet. The impact assessment must be made public at least fifteen days before the launch of the public enquiry (Article L. 123-10, Environmental Code).[44]

In terms of content, the impact assessment will review, *inter alia*, the quality of the project as regards its integration into the landscape, its effects on fishing resources, fauna, flora, its effects on ship traffic as well as the potential conflicts with other uses in the area. To that respect, certain areas are already protected under the regime of natural reserve or park, Natura 2000, heritage sites (zones de protection du patrimoine architectural, urbain et paysager, ZPPAUP). Ministry for the Ecological and Inclusive Transition has published a Guidance document for the completion of the EIA for offshore wind energy projects.[45]

In addition, the Law No. 2016-1087 of August 8, 2016, on the Recovery of Biodiversity, Nature and Landscapes requires the constitution of a financial guarantee as well as an annual fee for the benefit of the French Biodiversity Agency.

Finally, in the perspective of further development of economic activities at sea, the French authorities aim to establishing a common knowledge basis. As part of the tendering procedure, and in order to provide joint information to all interested actors, the French Energy Minister announced that studies on the wind, wave, depth and soil composition would be made public by institutions before the final submission of tenders, in order to give manufacturers the chance to consider risks and refine their proposals.

[3] Strategic Environmental Assessment (SEA) and Siting Related Planning

France is applying both international and European legislation in the domain. In particular, the provisions of the Directive 2001/42/EC on the assessment of the effects

44. On public participation scheme, *see* Section §3.03[A][6] below.
45. *Guide d'évaluation des impacts sur l'environnement des parcs éoliens en mer* (2017). *See* previous guidance documents: *Guide de l'étude d'impact sur l'environnement des parcs éoliens* (2010); Etude méthodologique des impacts environnementaux et socio-économiques des énergies marines renouvelables (2012).

of certain plans and programs on the environment is transposed into the Environmental Code. The latest amendments made to the consultation procedure for marine planning intend to comply with the requirements of the Directive.

[4] Planning the Legal Regime: Marine Planning and/or Land Planning

For a long time, France has been following, as in other areas, a centralized approach to marine spatial plan. For example, until 2005, the central government had sole competence on marine spatial plans. Since 2005, there has been a shift and local authorities are now also allowed to initiate and develop marine spatial plans. They can resort to the so-called SCOTs, "schéma de cohérence territoriale" or territorial coherence schemes that can unite several local coastal authorities (SCOT littoral). Today, consultation processes include more and more local interests in order to solve conflicts and find interaction in an early phase.

The increasing development of activities at sea has created a need for better assessment of both potentials and risks, and for better coordination of procedures.

Concerns around a better planning of activities and interests at sea were raised along the Environmental Roundtable (Grenelle II) and the Grenelle of the Sea. In the aftermath of the two Grenelle processes, several important initiatives took place: the creation of a National Council for the Sea and Coastlines (*Conseil national mer et littoraux*, CNML), the adoption of a National Strategy for the Sea and Coastlines, and organization of local consultation rounds for the identification of favorable sites for offshore wind development.

The CNML was established in January 2013 and is a new body created by the Law Grenelle II of July 12, 2010. The objective of the CNML is to reinforce the coordination of public authorities' action in the coastal areas. Importantly, the CNML is in charge of implementing the National Strategy for the Sea and Coastlines (*stratégie nationale de la mer et des littoraux*, SNML), which has been published on February 24, 2017.[46] The SNML is a central document which has to be updated every six years.

The launch of the first tenders for offshore wind has been a learning process. It has notably revealed the need for further coordination of activities at sea and a better marine space planning. Today, rules for marine planning are defined in the Environmental Code (Article L. 219-5-1). The Préfet and the Maritime Préfet play a central role in the process of identifying suitable location for offshore wind.

Broader consultation rounds took place as a new tool to identify the most suitable maritime areas for the development of offshore wind. The Grenelle II provided for a consultation forum for each maritime façade, with a particular focus on offshore wind energy potential. In 2009, when the Grenelle of the Sea took place, the government requested the coordinating Préfets to identify favorable sites and carry out consultation meetings with the goal to deliver an action plan by the end of the year 2010. On each coast, the process was ruled by a state authority (Maritime Préfet, regional Préfets).

46. Decree No. 2017-222 of February 23, 2017 on a national strategy for the sea and the coast, entered into force on February 25, 2017.

The use of local consultation as a tool for identifying marine areas suitable to offshore wind energy further developed. Lately, the consultation efforts have been accompanied by an effort to build the knowledge basis. Among the documents contributing to a better mapping of offshore wind energy projects in relation to marine space planning are:

- Study on the techno-economic potential of the zones suited for offshore wind energy installations, realized by the CEREMA on the basis of technical information provided by the industry.
- The French TSO, RTE, has conducted a feasibility study on grid connection for offshore wind.

On the basis of those studies, the government launched new initiatives aimed at identifying suitable zones for offshore wind energy projects. In December 2014, consultations with local stakeholders were launched on the basis of the two above-mentioned studies by CEREMA and RTE. The Préfet and the Maritime Préfet are responsible for coordinating this consultation and planning process. They are assisted by coastline marine councils.[47] Those consultations were organized by coastline and coordinated by the responsible Maritime Préfet.[48] The consultation process enabled building a consensus around new development zones for offshore wind. This consultation and mapping exercise will serve as a basis for the decision to launch call for tenders.

Besides recent initiatives, consolidation is needed. New mechanisms will be necessary to better coordinate activities in marine areas and to enable the development of offshore wind energy projects.

[5] International Law and/or European Law

France has signed and ratified UNCLOS Convention. Some claims on limits to maritime areas still remain in the following regions: Clipperton, Scattered Islands, New Caledonia, Mayotte, the EEZ in the Mediterranean, the sharing of the continental shelf in the Bay of Biscay, and the extension of the continental shelf of Saint-Pierre-et-Miquelon. The official position of the French government is that it will not submit the disputes relative to maritime limits to international jurisdictions. In the absence of a definitive settlement of a dispute, France retains sovereign rights over the areas thus delimited.[49]

France is a Party to the Barcelona convention on the Mediterranean and has ratified the Protocol on Integrated Coastal Zone Management (ICZM).

47. Coastline marine councils are advisory administrative bodies pursuant to Art. L. 2196-1 of the Environment Code.
48. Préfet de Haute Normandie for the maritime façade *Mer Manche du Nord*, Préfet des Pays de Loire for the maritime façade *Nord Atlantique Manche Ouest*, Préfet d'Aquitaine for the maritime façade *Sud Atlantique*, Préfet de PACA for the maritime façade *Méditerranée*.
49. *National strategy for the security of maritime areas*, adopted by the inter-ministerial sea committee on October 22, 2015, Prime Minister Office, French Republic.

[6] *Public Participation Scheme*

The French legislation contains several provisions ensuring public participation along the development of offshore wind energy projects.

First of all, the Environmental Code defines a procedure for public debate. A public debate must be conducted under the hospice of the French Commission of Public Debate each time a project increases the value of EUR 300 million.[50] The public debate is organized on an electronic platform.[51] The timeframe for completing a public debate is between three and four months.

Then, and in parallel, the legislation requires the organization of public enquiry at the local level. Both the award of the maritime domain concession and the water resource protection license require the completion of a public enquiry. In accordance with the Environmental Code, the two public enquiries can be conducted jointly.[52] The final report of the public enquiry is submitted to the Préfet of Department for consideration. The timeframe for completing a public enquiry is two months. Starting with the third tendering round for offshore wind bottom-fixed installations in 2013, the methodology for elaborating the tenders and assessing the projects include a consultation phase with stakeholders, including the local population. Many wind energy projects in France, both onshore and offshore, face strong opposition from the population. An emphasis on early consultation procedure aims to lower the level of conflict. In addition, ahead of the launch of any tender procedure, the Préfets (Maritime Préfet and Préfet of the region) are required to conduct a local consultation in order to precise the eligible area and ensure that the project has the support of the local population.

Third, after the adoption of Decree No. 2016-687 of May 27, 2016, the threshold for informing the public (*obligation de publicité*) about the project has been set to a generation capacity of 500 MW (Article R. 311-6, Energy Code).

[a] *Financial Participation*

Although financial participation in the form of equity, lending, energy cooperatives is an attractive participation model in France and encounter a growing success,[53] it is not yet applied for big offshore wind parks.

50. Article R. 121-2, Environmental Code.
51. Website of the French Commission of Public Debate https://www.debatpublic.fr/.
52. Article L. 123-6 and R. 123-7, Environmental Code.
53. Article L. 311-10-1, Energy Code, opens for equity participation in local renewable energy projects.

[7] Grid

RTE is the French TSO and is in a situation of natural monopoly for the electricity transmission network. This entails that offshore wind energy operators will need to connect their installation to the network operated by RTE and enter into a contract with it.

In the framework of the competitive dialogue under the latest tendering procedures, certain aspects of the connection agreement to RTEs are part of the negotiations. This is for example the case of connection costs. Similarly, even if the Energy Code provides that the applicant (electricity generator) can execute itself the connection work and cover the related costs,[54] those terms are determined as part of the tendering procedures. CRE approves the standard connection agreements.

In addition to the connection agreement, several addition agreements will need to be entered into between the wind energy producer and RTE, such as a grid access agreement (including grid access conditions, balancing responsibility, etc.), a testing agreement, an operating agreement (the testing agreement taking end), and a performance agreement.

If grid connection requires grid extension - which it is in most cases - RTE is required to get the necessary permits, such as maritime public domain concession,[55] but can benefit from a declaration of public utility.[56] Impact study and public enquiry requirements will apply.

RTE have received the mission to performing studies assessing the need for upgrading the transmission infrastructures to enable offshore wind energy generation development. RTE, in consultation with the DSOs, elaborate the regional scheme for grid connection of renewable energies (Schéma Régional de Raccordement aux Réseaux des Energies Renouvelables, S3REnR) (Article L. 321-7, Energy Code) which must be approved at the latest six months after the adoption of the regional plan for climate, air and energy (schéma régional du climat, de l'air et de l'énergie). These regional connection plans are considered as a means of reaching the renewable targets. Meanwhile, if the project is awarded through a competitive tendering procedures applied, the terms of the tendering procedures will prevail and not the one of the S3REnR. If the project is developed outside a competitive tendering procedure, the S3REnR applies.

In February 2016, the government has adopted a new Decree which aims at reducing the grid connection deadlines. It implements Article 105 of the Law on Energy transition for Green Growth. In addition, the legislation provides for compensation to the renewable energy generators in case of connection delays.[57] The connection deadline for electricity generation facilities based on renewables and of an installed capacity of more than 3 kVA is of eighteen months (Article L. 342-3, Energy Code).[58] As

54. Article L. 342-2, Energy Code.
55. Article R. 2124-1 to R. 2124-12, *General Public Entities Property Code.*
56. Decree No. 2017-81 dated January 26, 2017 regarding the environmental authorization; Decree No. 2017-82 dated January 26, 2017 regarding the environmental authorization.
57. Law of February 24, 2017 on renewable energies and self-consumption.
58. Article D. 342-4-1 to D. 342-4-6, Energy Code.

the connection of offshore wind energy facilities are technically demanding, the legislation provides a possible derogation to the eighteen-month deadline, subject to contractual agreement.[59] The compensation is to be paid by the transmission system operator, under conditions set by the Energy Code or the specific tender.[60]

[8] Other Concerns and Legal Regime

As in other countries, the development of offshore wind energy installations, whether they be bottom-fixed or floating, can enter into conflict with a series of other activities and interests. In France, the most common ones are: military activities, fishing, freight/passenger transport, tourism/recreation, sailing, nature protection and heritage sites. Only some of them are covered below.

[a] Military Activities

Access to marine areas for the development of offshore wind energy generation facilities are limited to a large extent by military activities due to military radars and training activities, flying zones. According to FEE, 47% of the total Metropolitan territory cannot be access due to military restrictions (both onshore and offshore). The 2015 Energy transition Law for Green Growth (Article 141) foresees the adoption of specific rules for the coexistence of renewable energy generation and military activities, but those have not been adopted yet.

[b] Habitat and Species Protection

Many attractive zones for offshore wind projects are located in protected areas. Some of the areas tendered for bottom-fixed installations are located in Natura 2000 sites.[61] Solutions for the integration of biodiversity and environmental protection concerns must be presented by the tenderers in their bid.

[c] Compensation Mechanism: Offshore Wind Tax

Local opposition has slowed down the development of wind energy in France, both onshore and offshore. A compensation mechanism has been developed over time to accommodate such opposition. The operator of offshore wind energy generation facilities is required to pay a special tax, when their installations are located in territorial waters or internal waters. The offshore wind tax must be paid annually, and is proportional to the MW capacity of the installation, with a fixed sum per MW of installed capacity (currently EUR 15,471).[62] The money collected through the tax is

59. Decree No. 2017-628 of April 26, 2017.
60. Article R. 342-4-10, Energy Code.
61. Such was the case for the third tendering round, launched in 2016 close to Dunkerque.
62. Article 1519 B and C, General Tax Code.

redistributed to local municipalities and "users of the sea": local municipalities from which the offshore wind farms are visible, based on the distance to the installations (50%); a national fund for compensation of wind energy development, and which support the development of sustainable exploitation of fish stocks (35%); and other projects for the sustainable development of other coastal activities (leisure, tourism, aquaculture).[63] The tax is conceived as an incentive to build offshore wind farms beyond the territorial sea.

§3.04 CHALLENGES AND SOLUTIONS

[A] Institutional Design: Challenges and Solutions

The institutional framework in place in France does not raise particular challenge. It is based on a centralized model, where the Ministry for the Ecological and Inclusive Transition plays a central role, in accordance with the competences conferred to it by the law.[64] The Préfet and the Maritime Préfet are both representing the state when awarding the concessions in their respective area of competence. The central role played by the Ministry and the Préfets is balanced by one of the CRE, which is, among other things, in charge of coordinating the procedures for competitive tendering in the sector of offshore wind. Relying on competitive bidding with negotiations enables solving a series of issues related to local opposition to wind projects and lack of competitiveness in the sector. The views of local authorities, in particular coastal municipalities, are better integrated as part of the hearing processes conducted both ahead of the launch of the tendering procedure and during the impact assessment. The sector has suffered since the start from a lack of competition between actors, and the establishment of a consultative process is also intended to preserve a higher degree of competition within the sector. In conclusion, although the procedure for developing offshore wind projects remains under the control of the state, either directly or indirectly via the Préfets, recent amendments to the legislation have enabled a better integrated of local interests.

[B] Incentives: Challenges and Solutions

The support regime for offshore wind has for a long time suffered from a lack of clarity and a high degree of complexity. Today, offshore wind projects can be developed both within and outside the competitive tendering procedure, according to the law. In practice, all projects have been subject to the launch of a competitive tendering procedure which has been replaced since 2013 by a competitive dialogue-procedure.

Relying on competitive tendering procedures is a requirement from the European Commission state aid guidelines from energy and environmental protection

63. *Ibid.*
64. Decree No. 2017-1071 of May 24, 2017 concerning the competences given to the Minister in charge of Ecological and Inclusive Transition.

(2014–2020). France is therefore complying with that requirement and is progressively aligning its support schemes onto EU requirements.

A remaining concern is the alleged preference given to national manufacturer in the award of the tenders. Promoting the development of a national industry in the offshore wind supply chain is a stated objective for the French government. For example, for some tenders, the turbines were manufactured in France (Le Havre), put together in French harbors (Boulogne-sur-mer, Saint-Nazaire, Brest), and maintenance services were provided by French companies. Local content requirements may contradict WTO rules and will need to be removed in most circumstances.

[C] Regulations: Challenges and Solutions

One of the reasons for the slow progress still is the high level of complexity of the permitting procedures and the lack of streamlining of the different parallel procedures. Projects are put in line waiting for grid connection, both onshore and offshore. Complaints against concessions, permits, and related judicial proceedings have also been a major source of delays.

French authorities, aware of those difficulties and keen to promote offshore wind, have recently revised the applicable regulatory framework. This must be seen as a first big step. The adoption of qualitative and quantitative targets has clarified the level of ambition of the government and has provided a higher degree of certainty for investors, developers and the national industry. Other positive changes are: the simplification of the permitting procedure and the establishment of the single environmental authorization; a better consultation process as part of the identification of potential zones for offshore wind farms and the national coastline strategy; and the adoption of a compensation mechanisms in favor of coastal municipalities. The one-stop-shop approach promoted by European legislation remains a goal.

§3.05 CONCLUSION

While renewable energy sources have been part of the French energy mix for a long time in the form of hydropower, wind power has retained a modest share in it. Onshore wind is progressively increasing, but offshore wind has still not developed. France has clear ambitions in the sector, both for bottom-fixed and floating installations. A late start has been the occasion of learning from the experiences gained by onshore wind in France and offshore wind abroad. France has recently amended its legislation to enable a quicker and larger development of offshore wind, in closer cooperation with industrial and local actors. This legislation builds on the principles of integrated coastal management and must develop within the framework of international law and EU environmental and state aid law requirements.

CHAPTER 4

Legal Framework to Develop Offshore Wind Power in Italy

Sandra Cassotta & Ulla Steen[*]

§4.01 INTRODUCTION

The full development of marine renewable energies, especially the offshore wind power sector, represents a huge opportunity for Italy which is surrounded by more than 8,000 km of coastal areas. The evolution in terms of growth is dependent of the National Energy Strategy ("Strategia Energetica Nationale" abbreviated as "SEN") with the goal of reaching the European Union (EU) objectives.

This chapter focuses on the regulatory framework applying to the offshore wind energy power sector in Italy and its interactions with the EU sources of law and policy objectives. According to the 20-20-20 strategy and the latest new EU targets, Italy should increase its share of energy production to 20% by 2020. Good results and investments have been achieved on land and in the onshore wind power sector. The Renewable Energy Progress Report (COM (2013) 175 shows particularly that in 2010 Italy already achieved its 2011/2012 interim targets laid down in Directive 2009/28/EC, reaching 10.4 share of renewable energy in the final energy consumption. Nevertheless, what is the situation with regards to offshore wind power sector in Italy? Why is there not a single offshore wind energy farm operating yet in Italy even though some of these projects passed the environmental impact assessment (EIA)?

The aim of this chapter is not only to provide an update of the current Italian regulatory framework but also of the past evolution and future trends of the sector under examination in order to provide a response as to why the sector is not operative.

* The authors would like to thank Fabrizio Penna, Senior Expert in the Cabinet Office of the Italian Ministry of the Environment and Protection of Land and Sea, Rome (Italy) and Associate Professor, Stefano Fanetti, University of Insubria, Como (Italy) for kindly provide some advice material and data.

The research question is to identify the legal and political challenges and barriers of the offshore wind power sector in Italy by examining the trends for increasing the market exposures of wind generators, and support schemes, such as feed-in premiums and competitive bidding procedures in order to address the most important concerns for investment which needs a certain stability of regulatory framework. The main problem in this sector is the existence of a cumbersome permit and licensing procedure and a very confused and heavy "authorization process" for operators willing to open an offshore wind energy farm. A clear mechanism of support is lacking, too many subjects are involved in the granting of permits, and it is not clear – "who licences what". This situation is aggravated by a chaotic and fragmented regulatory framework, and an uncertain distribution of competences between state and regions. This chapter also provides an analysis of the current national resistance against this sector in Italy by critically examining the normative framework and a seminal case law representative of fragmentation and legal uncertainty. This chapter concludes by identifying possible solutions at policy and regulatory levels and makes recommendations on how to overcome the difficulties in gaining permits in this sector in order to establish and run, thus allowing Italy to take profit of this great potential which up to now has been completely unexploited.

Even if the renewable energy sector in Italy continues to grow, there are currently no offshore wind energy farms in place yet.[1] The renewable energy sector has grown especially after the impulse of the EU strategy 20–20–20 of the EU and the huge economic incentives from the Italian government. The main problem is therefore not on the level of incentives, which have been quite high, but the absence of a national energy policy for a long time as well as a chaotic distribution of competence between state and regions in the renewable energy sectors. The evolution in terms of growth is now depending on the very recent SEN with the objective to reach the EU objectives but the past neglect or delay in designing this strategy, and this is a heavy burden and cause of barriers in the current development of this sector. The policy and legal framework of the offshore wind power sector is faced with the need to find and balance between the running of new wind farm and the needs of the society. The rapid development of this sector, in the future, will bring key benefits for the society especially in terms of reduction of Green Gas Emissions (GHG), and will favor the labor market, technologic development, and exports.

Nevertheless, the legal and policy framework applying to the sector must occur in harmony with important environmental and socioeconomic needs in order to guarantee sustainable development and the acceptance of public opinion to avoid what is called the "Not In My Backyard" (NIMBY) syndrome occurring when local governments and political parties oppose and protest against renewable energy installations.

1. Nevertheless, according to a latest update, at the time of the present writing, it is expected that the first offshore farm will be located in front of Taranto harbor in the Apulia region (Southern of Italy) in a water depths of 4–18 meters. Delivery and installation of turbines is currently planned for summer 2018, while their commissioning is expected to take place in autumn 2018. *See* more at http://www.offshorewind.biz/2017/06/14/senvion-turbines-for-first-offshore-wind-farm-in-the-mediterranean/?utm_source=emark&utm_medium=email&utm_campaign=daily-update-offshore-wind-2017-06-15&uid=14867.

In order to permit the Italian offshore wind energy power sector to achieve its role and finally to make use of its huge potential, there are some regulatory obstacles to overcome. A better and simpler regulatory licensing process and a more efficient access to electric grids are fundamental. There is a need to eliminate administrative barriers compared to the traditional renewable energy sources (RES), and this is the key problem in the Italian offshore wind power energy sector. In addition, there are some concerns as to the possible impact of offshore farms. In that respect, the offshore wind power sector must be respectful of the EU environmental legislation, which offers a very detailed legal and policy framework common to all Member States of the EU, in order to find solutions when problems of conflicts between national practices and the protection of the environment may arise. The mechanism of the EU sources of law and policy guarantees that the development of offshore wind power energy occurs in a sustainable way and reduces to a minimum its impact on the environment. In order to achieve sustainable development objectives in this sector and limit the potential impact of offshore wind farms on nature, it is fundamental to understand how and which type of legal duties an operator is supposed to respect in the phase of planning and management of offshore wind energy installation. This will be analyzed in the next paragraphs.

The geographical distribution of the current proposed offshore wind energy farms is mostly concentrated in the southern of Italy and also in the two islands: Sicily and Sardinia. The regions of interests are Puglia in the south but also Tuscany and Molise, situated in the central area of the "boot." Most of the potential is on "deep waters" rather than "transitional waters" or "shallow waters."[2] With regards to offshore wind energy farms, fifteen projects were presented to the government for approval in 2006–2013, but only two appear to have passed the stage of approval: one in the Gulf of Taranto and another opposite to the Gela coast.[3] Actually, there is no single offshore wind farm operating yet in Italy at the present time of writing even though some of the projects have passed the EIA procedure.[4] The project in the Gulf of Taranto provides a capacity of 30 MW with ten turbines. The project opposite to the Gela coast provides for a plant of 137 MW, with thirty-eight turbines and cables.

Hence, one wonders what is the barrier in Italy for the offshore wind farm to be rendered operative? In theory, in Italy we find all the conditions to make this sector operative as the country presents huge potential. The Italian Wind Energy Association, ANEV (Associazione Nazionale Energia del Vento) estimates the offshore wind energy of 2,500 MW is able to satisfy the needs of 1.9 millions of families. In addition, in line with the EU legislations, the SEN estimated that offshore wind farms would reach the

2. Marchisio, A., "APER" *Italian Renewable Energy Association* – Seanergy 2020.
3. Giugno, S., *"Eolico offshore, perché l'Italia non ha nemmento un impianto?"*, La Stampa, June 24, 2015, at http://www.lastampa.it 2015/06/24/ scienza/ambiente/focus/eolico-offshore.
4. Giugno, S., *"Eolico offshore, perché l'Italia non ha nemmento un impianto?"*, La Stampa, June 24, 2015, at http://www.lastampa.it 2015/06/24/ scienza/ambiente/focus/eolico-offshore; Marchisio, A., "APER" *Italian Renewable Energy Association* – Seanergy 2020; Legambiente "*Trivelle SI, Eolico offshore NO?" Da Taranto a Termoli, da Gela a Manfredonia tutte le barriere all'eolico in mare e il via libera alle trivelle,*" July 30, 2014.

objective of 100 MW, which were supposed to be installed in 2013 and which should produce a 680 MW in 2020.

The Renewable Energy Directive 2009/28/EC,[5] hereafter, the "RES Directive" established a common framework for the promotion of energy by setting mandatory national targets in order to achieve at least a 20% renewable energy share in the final energy output by 2020. The RES Directive required each Member State to set out the sectorial targets by their National Renewable Energy Action Plans (NREAPs) by June 2010.

Also, each of these individual plans defined the "technology mix scenario," the trajectory to be followed and reform measures to overcome barriers and ensure the developing of renewable energy. Offshore wind energy has a very important role in this context of "mix scenarios."

Italy already achieved its 2011/2012 targets, laid down in the RES Directive, as mentioned before in this chapter. But these good results have depended largely on the high level of incentives that have benefitted renewable energy resources, as it will be shown in the next sections. However, the trend of recent years is toward the reduction of incentives which could have serious impact on Italy's ability to reach its 2020 objectives. The functioning of the mechanism of incentives is explained and given by the "Gestore dei Servizi Energetici" (the "GSE") which is a very important actor in Italy. The GSE (in English, the "Energy Service Manager"), abbreviated "ESM," is practically a state-owned company which promotes and supports RES.[6]

The aim of the ESM is to foster sustainable development by providing support for renewable electricity (RES-E)[7] generation and by taking actions to build awareness of environmentally efficient energy uses.[8]

With regards to the schemes for the promotion of offshore wind energy in Italy, the Italian law implemented by the RES Directive is the D.Lgs 28/2011[9] which by transposing the RES Directive has introduced significant changes to the incentive schemes of energy from renewable sources.

Nevertheless, the sector started to decline, especially in 2014 partly as a consequence of the decreased level of economic incentives, which had been very high in the

5. Directive 2009/28/EC of the European Parliament and of the Council of April 23, 2009 on the promotion of the use of energy from renewable energy sources and amending and subsequently repealing Directives 2001/77/EC and 2003/30/EC (RES Directive), Official Journal (OJ) June 5, 2009, L 140/16.
6. RES or FER are Renewable Energy Sources (Fonti da Energia Rinnovabile) which means renewable non-fossil energy sources (wind, solar, geothermal, wave, tidal, hydropower, biomass, landfill gas, sewage treatment plant gas and biogases).
7. RES-E means "RES-Electricity," and it is the electricity produced from renewable sources which means electricity produced by plants using only renewable energy sources, as well as the proportion of electricity produced from renewable energy sources in hybrid plants also using conventional energy sources and including renewable electricity used for filling storage systems, and excluding electricity produced as a result of storage system.
8. *See* more about the "Gestore dei Servizi Energetici" (GSE), translated in this chapter by "the Energy Service Manager" (ESM) at the official website: www.gse.it.
9. Decreto Legislativo March 03, 2011, N. 28 Supplemento Ordinario N. 81/L alla Gazzetta Ufficiale N. 81/L "*Attuazione della Direttiva 2009/28/CE sulla promozione dell'uso dell'energia da fonti rinnovabili, recante modifica e successive abrogazione delle direttive 2001/77/CE e 2003/30/CE.*"

past. This could have serious impact not only on this sector but also on Italy's ability to reach its 2020 objectives. For example, the decrease in the level of incentives has led to a breakdown of Italy's installations for the onshore wind power sector in 2014 (107 MW installed against a media of 800 in the past years).

§4.02 INSTITUTIONAL DESIGN

Currently, in Italy, onshore wind energy farms are thus operative but not the offshore ones. For all the offshore wind energy projects presented to the government for approval, there have always been problems in the licensing and permitting procedures, especially at the level of the authorization process even though several projects had their EIA approved. Several actors are involved in the authorization process, which sees administrative proceeding complicating the acceptance from different ministries, regions, municipalities, and local entities which often enter into conflicts.

In Italy, the absence of a national energy plan for a long time, even though a plan was finally adopted but very late as mentioned previously, is also due to the remarkable diversity and heterogeneity in geographical, economic and environmental terms.[10] These problems have contributed to a disjointed and fragmented regulatory framework with respect to RES in general, and in the offshore wind energy power sector, in particular.[11] Specifically, with regards to the offshore wind energy sector, the lack of transparency in the permitting and licensing procedure, the lack of mechanisms of support and the interference of many subjects in the involvement on permitting and licensing, have created barriers in the development of the sector, especially for investors.[12] In the past, like in the present, it is not clear "who licences what."

Renewable energy projects are "share competences" between the regional competence and the national competence.[13] Nevertheless, there are some differences and exceptions in case of who shall deliver the "Authorization Procedure" to open a power plant. In case of onshore and other renewable energy, the competency is in the hands of the regions but not in the case of offshore wind power plants where the competency has been completely centralized meaning that in case of offshore, the competent actor is the State and not the Regions.

Hence, in Italy, authorizations in the case of the offshore wind energy power sector (differently to on shore) are issued by: (1) the public maritime domain (navigation code) granted in concession by the Ministry of Transport through Port

10. Fanetti, S. & Pozzo, B., "*Subnational Resistance against Renewable Energy: The Case of Italy*", in "Renewable Energy Law in the EU - Legal Perspectives on Bottom-up Approach", Peeters M., Shoemerus, T., Edward Elgar, 2014, pp. 165-183.
11. Fraterrigo, C., "*Profili critici dello strumento dei piani energetici regionali: in particolare, l'esperienza della Regione Sicilia.*" Norma - Quotidiano d'informazione giuridica, 13 Dicembre 2011.
12. Giugno, S., "*Eolico Offshore, perché l'Italia non ha nemmento un impianto?*", La Stampa, June 24, 2015, at http://www.lastampa.it 2015/06/24/ scienza/ambiente/focus/eolico-offshore.
13. Under the current provisions of the third paragraph of the Art. 117 of the Italian Constitution, "*'national production, transport and distribution of energy' is a matter attributed to concurrent legislative competence: the State sets out the fundamental principles, while the Regions are responsible for the preparation of detailed regulations.*"

Authorities), (2) the "Autorizzazione Unica" (Single Authorization) is granted by the Ministry of Transport (after consultation with the Ministry of Economic Development and the Ministry of the Environment and Protection of the Land and Sea, and (3) must go through an EIA which is granted by the Ministry of Environment, after hearing the Ministry of Cultural Heritage.[14]

Due to the absence of a unitary national legal framework in the RES in Italy, Regions often have worked disjointed from the other Regions and often in a random way by sometimes adopting more stringent and severe legal provisions or sometimes the opposite. Very often, bureaucratic burdens were added instead of simplifying the authorization procedure.

This had repercussions also in the transposition of the RES Directive and it is visible in the manner in which D.Lgs 28/2011 has been poorly applied and can be considered as an example of low level of quality at normative level.

The Case of the Court of Justice of the European Union (CJEU) C-2/10 Azienda Agro-Zootecnica Franchini Sarl[15] can be considered also as a very good case explaining the conflicts between the State and the Region where the main battleground is precisely represented by the authorization process for renewable energy projects.[16] This case has reached the CJEU and concerned the prohibition of wind turbines in the light of nature protection, and it is based on a preliminary question from the Italian Court regarding the total prohibition of wind turbines in nature conservation areas as the Puglia Region, a region that wanted to take action against the installation of wind energy. The case represents an example of a regional government willing to impose a barrier to further renewable energy establishment.

The CJEU explicitly considers the proportionality principle and states that Article 13 of the RES Directive introduces this principle with regards to administrative procedures for the authorization of plants producing renewable energy. In this case, the national Italian Court had to answer the difficult question whether this ban for the sake of nature protection is indeed proportionate and necessary. This case explains therefore where the objective of environmental protection gets weighted against the objective of renewable energy production when an offshore wind energy project will apply for authorization in a protected nature area.

The entrenchment between substantive aspects, that can be tested in a court procedure, and the need for administrative authorities to provide adequate justification in case of substantive burden is stronger than ever.

The CJEU determined that Article 194(1) of the TFEU includes the requirement for energy policy to have due regard for the need to preserve and improve environment[17] and that measures prohibiting only the location of a wind turbine not intended

14. Marchisio, A., "*Offshore Wind Power in Italy*" Seanergy 2020 Project APER (Italian Renewable Energy Association).
15. Court of Justice, Case C-2/10, *Azienda Agro-Zootecnica Franchini Sarl, Eolica di Altamura Srl v. Region Puglia*, July 21, 2011.
16. Fanetti, S. & Pozzo, B., "*Subnational Resistance against Renewable Energy: The Case of Italy*", in "Renewable Energy Law in the -EU Legal Perspectives on Bottom-up Approach", Peeters M., Shoemerus, T., Edward Elgar, 2014, p. 167.
17. Case C-2/10, July 21, 2011, para. 50.

for a self-consumption on sites forming part of the Natura 2000 framework with the possibility of an exemption for wind turbines intended for self-consumption with a capacity not exceeding 20 KW, in view of its limited scope is not liable to jeopardize the EU objectives of developing new and renewable forms of energy.[18] The CJEU held also that the Habitat & Birds Directives do not preclude more stringent national protective measures which impose an absolute prohibition on the construction of wind turbines not intended for self-consumption within areas of Nature 2000.[19]

§4.03 INCENTIVES

The fixation of tariffs and other implementing measures have to be decided in Italy, by a Ministerial Decree, and after 2013 the fare to be received during twenty-five years is determined by a tendering scheme with a based price for offshore projects of EUR 165/MWh.[20]

In order to be admitted to a tender process, bidders have to offer a reduction over the base price between 2% and 30%. Until the present time, it is reminded that no commercial offshore wind farm exists in Italy yet.

The incentive schemes currently provided by the Italian legislator for offshore wind energy power are based on: (1) *feed-in tariff*; and (2) *tendering schemes*. According to the *feed-in tariff*, plants with an installed capacity not exceeding 200 KW started after December 31, 2000, as a result of new construction, renovation or improvement, have access to a comprehensive fee that allows them to receive an incentive of 0.30 per KWh for a period of fifteen years.

With regards to the tendering schemes, Italy planned 680 MW of offshore wind energy for 2020 (129 MW by 2014). However, due to the fact that no offshore capacity has already been deployed, no new capacity has been allocated by tenders yet.[21]

In order to understand the evolution of the regulatory framework on the authorization process for wind energy power plants installations, it is important to examine, the regulatory trend from 2000 till now. A clear overview will help both to learn from the past mistakes and to improve the future legal conditions on permitting and licensing.

Between 2000 and 2004, the objective of the rules was just to "start" the market. In order to do so, the Italian legislator has introduced a system of "Green Certificates" ("GC"), in Italian "Certificati Verdi" abbreviated with "CV" with the D.Lgs No. 79/1999. The GC are green certificates which are a tradable commodity providing that certain electricity is generated using RES.

18. Case C-2/10, July 21, 2011, para. 57.
19. Case C-2/10, July 21, 2011, para. 58.
20. Gonzalez J.S., et al., "*A Review of Regulatory Framework for Wind Energy in the European Union Countries: Current State and expected Development*", Renewable and Sustainable Energy Review 56 (2016) 5, p. 593.
21. Gonzalez J.S., et al., "*A Review of Regulatory Framework for Wind Energy in the European Union Countries: Current State and expected Development*", Renewable and Sustainable Energy Review 56 (2016) 5, p. 591.

The GC ensured the stability and transparency and a strong administrative simplification with the procedure of "Single Authorization" through the D.Lgs No. 387/03 as previously mentioned[22] which involve all the public institutions in a single process authorization.

Between 2005 and 2008 a set of measures were established which went in the direction of development: (1) the incentive period was extended by eight years (initially provided by D.Lgs 79/1999) to almost a double, fifteen years (with the Law 244/07), and (2) the introduction of compulsory withdrawal of the GC expired by the ESM (L. 244/07 and implementing the Ministerial Decree 12/18/08).

From mid-2009 to the present time, there is high uncertainty of the system of rules: the attitude of the legislature has changed, and the priority no longer seems to be the development of the market as in the period between 2000 and 2004 but rather to achieve expenditure restraint.

In 2013, the D.Lgs 28/11 has provided the end of the "GC system" and a period of transition to an auction system better defined as "tender system" for large plants and administered price for others. Now, the "GC system" has been substituted by the "tender scheme" in 2013.

§4.04 REGULATIONS

The legal framework is confused, and it is not possible for offshore wind power installations to be subject to the same Guidelines (called "Guide Linea") that apply for onshore or other renewable energy installations. This is in total contrast with the practice of operators that applies for permits in other EU countries, such as Denmark, Germany, Spain or France, where the operators operate under much clearer legal frameworks. There is legal uncertainty as to how to evaluate projects, how to identify the areas to be protected, and how to inform citizens because on the sea it is not possible to apply the same guidelines applicable onshore. In Italy, the absence of a clear regulatory framework leaves open the possibility for public entities to block an offshore wind energy project, even if located several kilometers far from the coast or opposite to another installation or for landscape reasons and without any analysis of the projects. During Renzi's term of office, the choices in terms of changes in the regulatory and policy framework especially to simplify the authorization process have concerned fossil fuel installations. With the Legislative Decree (D.Lgs) "Sblocca Italia,"[23] improvements have only touched the offshore oil drilling for oil and gas extraction. This was

22. *See* footnote 16.
23. In 2014 three Italian laws ("decreti") were signed by the ministers regulating the system of incentives of renewable energies: (1) decree MisE-MATT concerning voluntary incentives (so-called spalma-incentivi volontario, Art. 1, paras 3–6 of the DL145/2013) for the power sector from sources other than photovoltaic; (2) decree MisE concerning rules on mandatory incentives for photovoltaic installations with power higher than 200 KiloWatt (KW) for a duration of twenty years (so-called spalma incentivi obbligatorio pursuant Art. 26, para. 3 of the DL 91/2014 for large photovoltaic installations); and (3) decree MisE on the methods of allocations for the incentives on the photovoltaic sector by the GSE, which reimburses to renewable energy producers, each year, with 90% calculated on the basis of the effective production of the

well explained in a letter of complaint sent on February 2014 by a group of Italian operators which had had their projects literally blocked.[24] The letter was sent to the President of the Council, Renzi himself but never received a reply.

In case of regional competency, the Decree 387/2003[25] contains important principles in order to rationalize and simplify the authorization procedures of renewable energy plants. First, Article 12 of Decree 387/2003 states that works to build plants producing electricity fuelled by RES as well as associated structures are considered as "public utilities," thus they cannot be "deferred" and must be considered as "urgent."

Second, a special administrative permit called "Single Authorization" has been introduced. The permit has to be considered by the Regions (or by "delegated provinces") through a single process which implies the involvement of all concerned administrative authorities.[26]

The procedural scheme chosen by the legislator is the "Conferenza dei Servizi" (in English "Service Conference" abbreviated "S.C."), which allows simultaneous representation and analysis of all various public interests and actors involved in the authorization process. In addition, Article 12 of the Decree 387/2003 also established the time frame for procedures. Article 12 also deals with compensatory issues, holding that the single authorization cannot be subordinated to or provide for such a measure in favor of Regions or Provinces.

Of particular interest of this Article 12 is number 10 that defers the punctual definition of the authorization procedure for renewable plants to further "Guidelines." On the ground of these Guidelines, Regions could intervene in order to identify areas and sites that were considered not suitable for installation of specific plants. These Guidelines are not applicable to the offshore wind energy power sector.

After 2007, we assist in Italy in the offshore wind sector for the establishment of a process of recentralization of authorization's procedure which was not applied for the onshore.

Thus, it is very important to note that during the last years, some exceptions to regional competence in the authorization process were provided in the case of offshore wind energy power sector.

previous year. This system should be in force until December 31 of the current year except if costs exceed EUR 5.8 billion on the costs of citizens' electricity bills. With these new laws, the Italian renewable energy power sector market appears deadlocked and will remain as such unless a new dress is adopted the day after the December 31, 2016 (which is the date of expiration of the decree). Therefore, in the Italian case, the future will depend of which kind of laws regulating incentives for renewable energy installations will be decided and whether an improvement of the system of division of competences between states and regions with the need to recentralize the competences to the State is adopted, and whether the costs of bureaucracy render the new laws inoperable. The need to overcome this deadlocked situation in the field of legislative competence seems to be supported by the Renzi Cabinet.

24. Legambiente "'*Trivelle SI, Eolico offshore NO?' Da Taranto a Termoli, da Gela a Manfredonia tutte le barriere all'eolico in mare e il via libera alle trivelle*", July 30, 2014. This report contains a reproduction of the letter on p. 3.
25. Decreto Legislativo n. 387 del 29.12.2003. G.U. n. 25 of the January 31, 2004, Suppl. Ordinario n. 17.
26. Fanetti, S. & Pozzo, B., "*Subnational Resistance against Renewable Energy: The Case of Italy*", in "Renewable Energy Law in the EU – Legal Perspectives on Bottom-up Approach", Peeters M., Shoemerus, T., Edward Elgar, 2014, pp. 165–183.

First, Article 2, paragraph 158, letter C of the Law 244 of 2007[27] established that the authorization for the offshore installations be issued by the Ministry of Transport, the Ministry of Economic Development and the Ministry of Environment and Protection of Land and Sea. Second, an additional change in the law, which was introduced by Article 13 of the Decree 46/2014, has assigned to the Ministry of Economic Development the competence of authorization of renewable plants with total installed heat capacity equal to or greater to than 300 MW. These amendments may be viewed as part of the recentralization process of renewable energy governance that the Italian government is undergoing.[28]

Both in the case of regional competence and national competence in accordance with the Legislative Decree 152/2006, an "EIA" applies. This Legislative Decree approves the Code on the Environment which sets out the legislative framework applicable to all matters concerning environmental protection. The Code is composed of six parts. One of these parts concerns and defines and regulates the procedure of EIA (in Italian "Valutazione di Impatto Ambientale" abbreviated "VIA").

This means that wind power plants are linked to EIA procedures and "other procedures." For wind power plants exceeding 60 KW of power, wind power plants regardless of their sides are subject to Screening Procedure. If these plants exceed 60 KW and are partially within the system of Protected Natural Areas, they are subject to an EIA. The head of the EIA must ensure that the procedure for granting the Single Authorization Procedure is coordinated under the procedure of impact assessment.

For offshore wind power plants which are less than 60 KW, after the new Decree DL 91/2014 converted in Law 116/2014 amending the aforementioned D.Lgs 152/2006, it is also mandatory for small offshore power plants to be subjected to EIA.[29] These are also "other procedures" as mentioned in the previous section, that are requested for the installation of offshore energy power plants. These procedures other than the "Single Authorization" are, for example: Incidence Assessment (Valutazione di Incidenza), Declaration of Starting Activity (Dichiarazione di Inizio di Attività), "Certificate of no Impediment" (Nulla Osta for Landscape and Archeological and Flight Constrains) and an Integrate Environmental Authorization (Autorizzatione Integrata Ambientale).[30]

Tax regulations mechanism exists to reduce real tax amount exists. They are to reduce real tax amount to less than 0.4%. The deduction is valid for a maximum period

27. Legge n. 244/07, Dicembre 2007.
28. Fanetti, S. & Pozzo, B., "*Subnational Resistance against Renewable Energy: The Case of Italy*", in "Renewable Energy Law in the EU – Legal Perspectives on Bottom-up Approach", Peeters M., Shoemerus, T., Edward Elgar, 2014, p. 167, footnote 10.
29. Sezioni Unite Penali (Penal Sections) of the Corte di Cassazione in the recent judgment of April 13, 2016 n. 15453 that made the state of the art in the regulatory evolution in the field of EIA after the procedure of infraction (messa in mora) against Italy for not transposing correctly the Directive 2009/31/EC on Environmental Impact Assessment. For that purpose, the Italian legislator has enacted the Law 91/2014 converted by the Law 116/2014 that modifies the D.Lgs 152/2006 which provides that irrespectively from the thresholds established in the Allegato IV of the Codice dell'Ambiente (Environmental Code) all the installations must be submitted to an EIA.
30. Calabrese, G, "*L'Industria dell'Energia Eolica in Italia – Elementi Strutturali e Dinamiche Competitive*", 2012, Cacucci Editore, p. 158, footnote 65.

of five years from the date of installation of the plant.[31] This tax is determined at city council level. Also a tax regulation mechanism exists (reduction in value added tax). The reduced value added tax rate is 10% (instead of 20%).[32]

Grid connections have been set up and defined as "shallow cost approach."[33]

The shallow coast approach means that the plant developer bears the costs of equipment necessary to connect the generator to the nearest point on the already existing grid network. On the other hand, the grid owner will bear the cost of any reinforcement that would be necessary to integrate the new generator. There are some barriers in that sense as grid connection requires a long time and the grid is underdeveloped (as it will be explained in the next section).

It is worth noticing here that the reason why the grid connection issue is taken into consideration in this analysis is because the connection of offshore wind farms to the grid is actually very important. One of the most crucial preconditions for offshore wind power deployment is the existence of an adequate regulatory and economic framework of the connection of offshore wind plants to the onshore transmission grid. In the best case, planning and construction of one or several offshore wind plants go hand in hand with planning of the relevant cable connecting the offshore wind plant to shore.[34] In the worst case, large offshore wind plants are completed but are unable to commence operations whatsoever because of delayed grid connections. This causes large economic losses and represents barriers for investors.

On November 23, 2015 the Italian Regulatory Authority for Electricity Gas and Water (AEEGSI) adopted a resolution[35] that provides general conditions for the connection to the renewable energy plants of the national grid. The AEEGSI Resolution rules for the first time the connection procedures related to wind offshore plants built on Italian national waters. The resolutions states that "Terna" (translated in English as "the National Grid Operator"), subject to the prior consultation and approval by the

31. Article 1, C.6, l.a. L 244/07.
32. *See* at, http:// www.res-legal.eu/search by-country/italy/.
33. Grid issues are related to issues regarding grid connection, which is to say, the procedure and coast allocations and operation (priority use of the grid and balancing). The general procedure for grid connection in most European countries is such that after performing the basic technical projects of the wind farm, the plant developer sends the application to the operator. There are different approaches in the EU Member States for sharing costs of grid connection between producer and grid operators: (1) shallow cost approach (as described above in the text and which apply to the Italian case), (2) super shallow approach which means that the plant developer only have to bear the costs of the inner electrical infrastructure including plant substation, and (3) Expansion of the grid to the connection point and reinforcement which is borne by the grid operator, (4) Deep coast approach which means that the plant developers have to bear all connection costs, as well as any further reinforcement expenses that can arise as a consequence of integrating the generator in the electrical system, (5) Mixed shallow approach which is a model defined as "hybrid" of the deep coast and super-shallow approach. *See* more in Gonzalez J.S., et al., "*A Review of Regulatory Framework for Wind Energy in the European Union Countries: Current State and expected Development*", Renewable and Sustainable Energy Review 56 (2016).
34. Pira, R., Report "*EU and Regional Practices for Offshore Wind: Creating Synergies*", Foundation of Offshore Wind Energy, (November 2014) p. 21.
35. Resolution 558/2015 amending and updating AEEG Resolution 99/08.

AEEGSI, will specify within the Grid Code (Codice di Rete) the possible solutions for the connection of the offshore wind plants.[36]

In particular, the regulatory delays are due to bureaucratic issues and instability of incentives schemes, and these two impacts not only on time but also on costs. In addition, the costs of connecting the grid are very high such as for the construction of electrical substations and sometimes there are royalties to the municipalities involved in the initiative phase, and it should not be forgotten that wind farms in Italy are also subject to property taxation (around EUR 5,000/MW).[37]

Moreover, the grid is underdeveloped in Italy because in order to ensure the safety of the electrical system, an entity which is "Terna" can impart some limitations of production in a planned manner or in real time referred as "dispatching orders." With that respect, some Italian producers may present to the ESM a request to obtain the remuneration for lost generation.[38]

The objectives enshrined in the Treaty of Functioning of the European Union (TFEU) need always to be taken into consideration when it comes to the authorization of an offshore wind energy power installation including grid connections. Here lies, on one hand, the strong linkage between procedures and authorization, and on the other hand, the protection of the environment. The two often need to be literally "weighted" against the EU secondary sources of law protecting the environment, in particular in interaction with several crucial directives that will be considered in this section. The objectives enshrined in the TFEU framework that need to be taken into consideration when authorizing the construction of offshore wind farms and grid interconnection are: (1) Protecting the environment (for instance where offshore wind farm are constructed in nature-protected areas and combating climate change); (2) Achieving a competitive and secure internal energy market with a free movement of electricity; (3) Promoting the interconnection of energy networks and the development of trans-European networks in the area of energy infrastructure.

Member States of the EU in question need to consider whether the offshore wind power farm projects jeopardize other objectives under EU law, for example, the protection of biodiversity and in that respect, Member States can adopt stringent requirements that in a given case prevent the authorization of renewable energy project.

A concrete example of this situation is the CJEU Italian's Case C-2/10, Azienda Agro-Zootecnica Franchini Sarl previously mentioned where it was important to understand what objective the construction of an offshore wind farm or grid seeks to achieve and where objective and authorization requirements are connected too and

---

36. The resolution also delegates other grid operators to issue their own solutions and contractual conditions in order to carry out the connection of offshore wind plants.
37. *See* the APER Report, *Associazione Produttori Energia da Fonti Rinnovabili*, 2012.
38. The regions in Italy most affected by the phenomenon are in the south of Italy, in particular in Puglia and Campania where there is a high concentration of plants. In case the grid would have been sufficiently developed, production from wind plants in 2010 would have amounted to 9.606 GWh, plus 5% (480 GWh) compared to 9.126 GWh real. *See* APER Report, Associazione Produttori Energia da Fonti Rinnovabili, 2012.

balanced against the requirements in the RES Directive and other objectives contained in other directives protecting the biodiversity.

In particular, the Habitat & Birds Directives[39] are fundamental with the objective to protect the species in the EU and preserve natural habitat for certain specie at risk on Natura 2000.[40]

In particular, Article 6, paragraph 3 of the Habitat Directive provides that any plan or project likely to have a significant effect on the management of the site, either individually or in combination with other plans or project, is subject to an assessment of these implications for the site in view of the site's conservation objective. According to the CJEU, these directives do not prohibit all human activities within a Natural 2000 site but simply make authorization of such activity, conditional upon a prior EIA.[41]

Therefore, the authorization process for offshore wind energy farm is legally entrenched between the Habitat and Birds Directives, and the directives related to environmental and strategic impact assessment binds Member States' legislations to conduct impact assessment because it is linked to the issuance of authorizations to install power plants.

The EIA Directive is an EIA procedure in force since 1985, applying to a wide range of defined public and private projects, has been modified in 1997 (97/11/EC), 2003 (2003/35/EC) and 2009 (2009/31/EEC).[42] In Italy, the implementation of this Directive, in particular in the case of an offshore wind energy power plant, is an example of how complicated and numerous the considerations are concerning the authorization process. They have to be taken into consideration often become battle-field between state and regions, and the amount of documentation that is needed for the issuance of authorizations to set an offshore wind farm.[43] The procedures are numerous, as already analyzed and described in the previous section.[44]

In the process of the EIA, according to Articles 6, 7 and 8 of the EIA Directive, public shall be consulted which in the offshore wind power in Italy is translated as the participation of the public which can be consulted on environmental issues before the issuance of the decision of authorization. The results are therefore taken into consideration in the process of authorization.

39. Council Directive 92/43/EEC of May 21, 1992 on the Conservation of Natural Habitat and the Protection of Natural Habitats and of Wild Flora and Fauna, OJ L 206 of 22.07.1992 and Council Directive 79/409 EEC of the April 2, 1979 on the Conservation of Wild Birds, OJ L 103 of 25.04.1979.

40. Natura 2000 is a network of core breeding and resting sites for rare and threatened species and some rare natural habitat types which are protected in their own rights and covers twenty-eight EU Countries including land and sea. The aim of the network is to ensure the long-term survival of Europe's most valuable and threatened species and habitats, listed both under the Habitat & Birds Directives. *See* "Natura 2000 – Environment" at: ec.europa.eu.

41. *See* Court of Justice, Case C-2/10, *Azienda Agro-Zootecnica Franchini Sarl, Eolica di Altamura Srl v. Region Puglia*, July 21, 2011.

42. Directive 85/337 on the assessment of the effects of certain public and private projects on the environment, OJ 1985 L 175/40, as amended by Directive 97/11, OJ 1997 L 73/5 and Directive 2003/35, OJ 2003 L 156/17.

43. Calabrese, G, "*L'Industria dell'Energia Eolica in Italia – Elementi Strutturali e Dinamiche Competitive*", 2012, Cacucci Editore. p. 158, footnote 65.

44. *See* previous section.

The involvement of local communities in renewable energy plannings is not sufficient to deal with the increasing number of protests. According to an Italian Report of 2009,[45] protests are not dealing only with the offshore wind sector but also with different other sectors and evidence shows how a limited or nonexistent access for the public to the decision-making process had a negative impact on the realization of power plants.[46] According the Regional Law of 2007, public debate is possible and not compulsory for projects that present important environmental and social impact. Nevertheless, there is a need to improve the involvement of local communities in the choice of energy infrastructures to avoid form of oppositions as there is a lack of participation in the planning process due to the absence of correct information. The Strategic Environment Assessment Directive, better known as "SEA Directive," sets the rules for strategic environmental assessment of the effect of certain plans and programs.[47] Also, provisions related to EIA and SEA Directives are different compared to "Incident Assessment" (Valutazione di Incidenza) as mentioned in the previous section[48] which explains the amount of bureaucratic passages and documentation.

Without any doubt it would be useful for Italy to find other sources of inspirations from other countries, such as the Danish experience, for example, that provides authorizations procedures quite simplified which award the power of the installations and interesting incentives schemes. This country implemented specific support schemes or adapted to the remuneration level based on *feed-in premium* FiPs,[49] financing and tendering schemes.[50] These Danish measures include interesting support for the financing of the preliminary investigations by local wind turbines and adoption of local purchase by citizens and *premiums* based on the power of energy installations. The *feed-in premium* FiPs is based on: (1) call for tender, and (2) and open-door procedure.[51] For the coastal projects, a 20% share of ownership of the projects has to be offered to local residents or companies. Guarantee for loans taken out by local owners are provided by the Danish transmission system operator called "Energinet.dk." This could be also a solution in case of local opposition that delay or block

45. NIMBY Forum, V edizione (2009), Cantiere Italia. Quando lo sviluppo è una corsa a ostacoli, Aris.
46. Fanetti, S. & Pozzo, B., "Renewable Energy Law in the EU – Legal Perspectives on Bottom-up Approaches", Peters, M., Shoemerus, T., Edward Elgar, 2014, p. 178.
47. Directive 2001/42 on the assessment of the effects of certain plans and programmes on the environment, OJ 2001 L 197/30. An assessment under the EIA Directive is without prejudice to the requirements of the SEA Directive, Case C-295/10 Valciukiene and others, Judgment of September 22, 2011, Paragraph 59. *See* this point in Jan H. Jans & Vedder Hans H.B., "*European Environmental Law After Lisbon*", Europa Groningen Publishing (2012) p. 366, footnote 32.
48. *See* previous section.
49. *Feed in premium* FiP are defined as market dependent mechanism.
50. Olsen B.E., "*Regulatory Financial Obligations for Promoting Local Acceptance of Renewable Energy Projects*", Chapter 10, in "Renewable Energy Law in the EU – Legal Perspectives on Bottom-up Approach", Peeters M., Shoemerus, T., Edward Elgar, 2014.
51. In Denmark, in the call for tenders, the subsidy to be received in form of a sliding feed in premium is selected by a tender procedure. With regards to the open-door procedure, offshore wind farms under the feed in tariff premium of EUR 30 /MW h for 22,000 equivalent full hours plus EUR 3/MW h for covering the balancing costs. Gonzalez J.S., et al., "*A Review of Regulatory Framework for Wind Energy in the European Union Countries: Current State and expected Development*", Renewable and Sustainable Energy Review 56 (2016) 5, p. 593.

installations. The Development in the Danish Law has been introduced as a new measure to promote onshore and offshore areas. Specifically, the Danish Renewable Energy Act[52] has introduced specific measures: (1) local ownership approach which includes a fund support; and (2) a coownership scheme which imposes an obligation on wind energy developers to offer a minimum of 20% ownership of project to local citizens.[53]

§4.05 CONCLUSION

Even though Italy presents all the potentials to have a strong and prosperous future with offshore energy power, the future is black if the current regulatory and policy framework remains as is and might even lead to a progressive disruption of the current installations. There are numerous points of criticism that represent real barriers for current investments and discourage future operators. There are very strong delays in permitting procedures which are too complicated in part, due to the power of the Region to delay the conclusion of permitting procedures. In addition, the lack of information for operators not thoroughly educated in laws is a reality which furthers the NIMBY syndrome. From the examination of the evolution of the normative period between 2000 until days, it can be concluded that the normative transition in the offshore wind energy power sector in Italy has not been satisfactory. Often there are too many subjects in the decision-making which impede the objective to achieve a satisfactory deal with regards to the all interested parties and stockholders and their interests at stake. Operators must gather a huge amount of opinions and have dialogue with too many actors (natural and legal persons, organizations, institutions, Ministries, etc.). An improvement would be to change strategy with fewer actors and institutions involved where stakeholders would be aware of who or what really counts. A second improvement would be to change the logic of management and regulation as there are too many Ministries involved and too many high costs, for example, costs of auctions and costs of registers mechanism and grid connection mechanisms. In addition, there is a lack of competence from Port Authorities which are military bodies and often have no background or understanding of renewable energy law and policy. Nevertheless, they are still part of the process of authorization and sit at the board of the decision-making process in the authorization process.

In the light of this analysis, it is therefore recommended to (1) avoid letting the current situation affect the whole sector; and (2) modify the current incentive system by simplifying the current permit and licensing system establishing, for example a bonus for production increases, and (3) to find possible solutions to get free from this impasse by finding sources of inspiration from other countries when redesigning the regulatory framework. Countries such as Denmark or Germany which have much more

52. First renewable Energy Acts, Act No. 1392 of December 27, 2008 as replaced by Consolidated Act on Renewable Energies No. 1074 of November 8, 2011.
53. Olsen B.E., "*Regulatory Financial Obligations for Promoting Local Acceptance of Renewable Energy Projects*", Chapter 10, in "Renewable Energy Law in the EU – Legal Perspectives on Bottom-up Approach", Peeters M., Shoemerus, T., Edward Elgar, 2014, pp. 195–196.

targeted programs, simplified normative procedures, and not too long and burdensome authorization processes, could be the perfect source of inspiration to redesign Italy's offshore wind energy power sector regulatory framework.

CHAPTER 5
Legal Framework to Develop Offshore Wind Power in Norway

Catherine Banet[*]

§5.01 INTRODUCTION: POLICY BACKGROUND

Norwegian renewable energy policy is at a crossroad in many ways, influenced by both internal and external factors of technical, economical and regulatory nature. This situation is well illustrated by the current state of development of the offshore wind energy sector. Norway is facing a paradox of having a new regulatory framework in place enabling the construction and operation of offshore wind generation facilities, but being unable to develop a domestic market for offshore wind. While the political willingness is there as proven by the adoption of two recent parliamentary resolutions, Norway is still lacking a momentum for starting commercial offshore wind projects. There is consequently almost no experience with offshore wind in Norway, except some few demonstration projects. Meanwhile, Norwegian authorities and commercial actors see promising potential in the development and export of offshore wind technologies, in particular floating installations.

[A] Political Will to Develop Offshore Wind

Norway has a particular energy profile as nearly all its electricity generation is based on renewable energy sources, and dominantly hydropower. The share of renewable energy sources in the production mix is consequently much higher than in other

* The author would like to thank Daniel Willoch from the Norwegian Wind Energy Association (NORWEA) for his valuable comments on the text. The author remains sole responsible for the content of the chapter and the conclusions drawn.

European countries. Almost 96% of electricity generation is based on hydropower.[1] Among the other renewable energy sources of electricity production, wind power counts for less than 2% (1.7% in 2015).[2] Consequently, wind power is still taking a minimal part of the share of electricity generation mix. The operative commercial wind energy projects are all located onshore. Except one demonstration project in operation and another one planned, there is no offshore wind electricity generation as of 2017.

Current renewable energy policy in Norway is primarily driven by its commitments under the Renewable Energy Directive 2009/28/EC which has been incorporated into the European Economic Area (EEA) Agreement in 2011 and which legally binds Norway.[3] The negotiation for the incorporation of the Directive resulted in the adoption of a target of 67.5% share of renewable energy sources in the final gross energy consumption of Norway by 2020. To achieve this goal, Norway has established from January 1, 2012 a national green certificate scheme coupled with a joint green certificates market with Sweden. Based on a bilateral agreement, the two countries have agreed that, in the nine years from 2012 to 2020, new renewable electricity generation capacity of a total of 26.4 TWh shall be developed in total for Norway and Sweden jointly. The scheme is technology neutral, and both onshore and offshore wind is eligible. Norway is committed to finance 50% of the certificates, regardless of where the production is located in each of the two countries. This is the only concrete target in terms of new renewable electricity generation adopted and binding on Norway. Meanwhile, there is a consensus among Norwegian political parties for not extending the scheme after the current period (last issuance of certificates in 2020, last cancellation in 2035).

A second source of commitment is created – although more indirectly – by the international climate regime. The effects of that regime on renewable energy generation in Norway are more indirect that the EU target because of the particular energy generation profile of the country. The reduction of greenhouse gas (GHG) emissions result primarily in actions at the level of energy consumption within the offshore petroleum sector, industrial activities and, more recently, transport.

Norway ratified the UN Framework Convention on Climate Change (UNFCCC) on July 9, 1993 and the Kyoto Protocol on May 30, 2002 and became party to the Protocol when the latter entered into force on February 16, 2005. Under the second commitment period of the Kyoto Protocol (2013–2020), Norway agreed to limit its average annual emissions of GHG gases to 84% of emissions in 1990. It corresponds to a reduction of global GHG emissions equivalent to 30% of Norwegian emissions by 2020 compared to 1990. This is an ambitious target and raises question of compliance strategy as the country relies heavily on buying carbon credits to meet its own emission reduction goals.

1. In 2015, 95.8% of electricity generation in Norway was based on hydropower. Source: Statistics Norway (SSB), November 2016.
2. Source: SSB, November 2016.
3. European Economic Area (EEA) Agreement, Annex IV – Energy, point 41, inserted by Decision No. 162/2011, entered into force December 20, 2011.

Following the adoption of the European Union (EU) Climate and Energy Strategy for 2030 and in advance of the 21st Conference of the Parties to the UNFCCC (COP21/CMP11) held in Paris in December 2015,[4] the Norwegian government has slightly reviewed its own climate policy. In its 2015 White Paper on Climate Policy, the Norwegian government suggests that Norway by 2030 will reduce GHG emissions by at least 40% compared to the 1990 level. To fulfill this target, the government seeks a close cooperation with the EU. EU climate and energy policy falls outside the scope of application of the EEA Agreement. Meanwhile, Norway is following closely EU policies in that domain and the suggested Norwegian target is in line with the EU climate targets. Norway has also announced its intention to enter into a bilateral agreement with the EU for the joint fulfillment of the climate targets, based on the EU framework for climate policies.

Norway has signed the 2015 Paris Agreement on April 22, 2016 and ratified it on June 20, 2016. Norway's first Nationally Determined Contribution (NDC) mentions renewable energy as one of the measures that the country intends to promote in order to reduce its GHG emissions.

Finally, the Norwegian Parliament enacted a new Climate Act which enters into force on January 1, 2018.[5] The law defines binding GHG emission targets for 2020, 2030 and 2050. It also sets a series of carbon budgets for the government and may influence the further development of new renewable energy sources in Norway.

[B] Rationale for Moving from Onshore to Offshore: National Policy

The move from onshore to offshore wind is not a straightforward process in Norway. To become a reality, it needs further technological innovation, costs reduction and a clear implementation strategy from the government. Meanwhile, it has a strong export potential, is seen as a long-term business opportunities and could play a supplementary role for electricity supply at domestic and European level.

The first motivation comes from the ambition of developing a new export industry. Norway has a strong competence within the offshore industry based on both the shipping and oil and gas sectors. Many demonstration projects for offshore wind installations are based on the competence acquired by the offshore and subsea construction industries, and are developed by the very same companies (e.g., Statoil ASA, Fred Olsen Renewables AS). Norwegian companies are already exporting technology and know-how to offshore wind developers all over Europe and look increasingly at the U.S. and Asian markets. Among them are Fred Olsen Windcarrier (installation of wind turbines with specialized carriers) and Owec Tower (jackets, steel fundaments for offshore wind mills). A recent report ordered by Export Credit Norway estimated that Norwegian suppliers accounted for approximately 5% of deliveries to offshore wind projects since 2010. However, in some offshore wind sub-segments, the

4. The EU has decided that the emissions that are covered by the European Emission Trading Scheme (EU ETS) are to be reduced by 43% in 2030 compared to 2005. In the sectors not covered by the EU ETS, the EU will reduce emissions by 30% compared to 2005.
5. Climate Act (*Lov om klimamål*) of June 16, 2017.

Norwegian presence is considerably higher, such as within cable supply and installation, installation vessels, marine operations and operations and maintenance (O&M) services.[6] In conclusion, the Norwegian suppliers still takes a limited share of the global offshore wind market, but are expected to extend their activities abroad, making use of the competences of the shipping and offshore industry and relying on some innovative technological solutions.[7]

Although most Norwegian actors involved in the sector are working primarily on the development of demonstration projects and looking at the export markets, some few others aim to reinforce their operatorship competences. Statoil and Statkraft (even sometimes together)[8] are involved in offshore wind projects abroad. Statoil's recent strategy has been to reinforce its position as operator in different projects in the UK and Germany in order to improve efficiency and increase competitiveness across projects (Sheringham Shoal, Dudgeon and Dogger Bank offshore wind farms in the UK and the forthcoming Arkona wind farm in Germany).

In conclusion, Norwegian actors aim to both gain operatorship experience from commercial markets abroad and export products and know-how to *inter alia* the United Kingdom, France, Germany, the United States and fast growing Asian markets.

By contrast, this competence is not used for developing offshore wind projects along the Norwegian coastline. There is no domestic market for offshore wind power at the moment. Moving offshore in Norway is challenging due to the non-homogenous seabed geology and technical challenges. Norway, with its long coastline (25,148 km), benefits from very good wind resources. Meanwhile, strong winds, high waves, seabed typology and water depth represent important natural challenges. Water depth increases material costs for the foundation work in fix-bottom installation. This naturally limits the number of available areas. In its 2010 study, the inter-directorate group in charge of proposing the Strategic Environmental Assessment (SEA) agenda for offshore wind considered depths to 70 m as being feasible for fixed-bottom wind power.[9]

In that context, floating turbines can represent a valid alternative, with more flexible requirements as to depth. In its 2010 study, the inter-directorate group retained depths between 120 and 400 m are feasible for floating installations.[10] Statoil has come far with its floating windmill, called Hywind (demonstration project outside Karmøy), as well as a pilot park in Scotland and the United States.

Onshore wind may enter in conflicts with many interests, would they be related to economic activities onshore or along the coasts or to landscape protection. Local population's opposition has proved to be strong. Moving offshore may help solving certain of those conflicts. Therefore, current focus is on collaborative projects where

6. Report by MAKE Consulting for INTPOW, Export Credit Norway and Greater Stavanger, *Norwegian Opportunities in Offshore Wind*, 2017.
7. Energiogklima Stiftelse and Norsk Industri,*Et hav av muligheter*, Norsk Klimastiftelse / Norwegian Climate Foundation, 2012.
8. In 2009, Statkraft and Statoil joined forces to build the 317 MW Sheringham Shoal Offshore Wind Farm (88 wind turbines) off the coast of Norfolk, England. Sheringham Shoal is now owned by Statoil, Statkraft and The Green Investment Bank through joint-venture company Scira Offshore Energy Limited.
9. 2010 Study, Forslag til utredningsarbeid.
10. 2010 Study, Forslag til utredningsarbeid.

offshore wind can coexist and contribute to the development of other resources. This can also solve the question of conflicts in area use.

Providing a concrete and promising area of collaboration, several test projects already envisage using offshore wind to supply electricity to offshore petroleum platforms. Offshore wind was also envisaged as a source of electricity supply to petroleum platform as part of the Siragrunnen project. Electricity from offshore wind can be used to supply not only platforms but also subsea installations, the latter being extremely energy-intensive (for the moment supplied by electricity cables from land or from other existing platforms).

Another conflict area which could result in collaboration is the one between offshore wind and fish farms. Norwegian research institutions and commercial actors are working on several research projects, such as Gwind, aimed at combining offshore wind electricity generation and fish farming. It follows an international trend towards more off-grid installations. It can take the form of vertical wind turbine supplying electricity to the salmon farm. The biggest technical challenge at the moment is the storage capacity to answer the needs of the installation at all time (stable electricity generation is essential to maintain good hygiene conditions for fish farming).

In all those promising applications of offshore wind, there is still a knowledge gap which needs to be filled. Although the public authorities have conducted an assessment work in 2010 and 2012, and can refer to European data (e.g., costs, efficiency), the national conditions and impacts on birds, fish and marine mammals differ from other countries and requires local assessments. In addition, technical innovation evolves very quickly, which makes it difficult to rely on old figures as a basis for sound decisions. The 2010 study concluded on the obvious need for data collection and preliminary studies.

[1] Policy Setting

At the general level, the government places great emphasis on increasing supply of renewable energy. During the period 2006–2011, new facilities with a total production capacity equaling about 5.3 TWh became operational.[11] However, wind energy remains a marginal contributor to the energy mix. Therefore, the industry representatives have been critical to the lack of initiative of the government in the successive energy policy papers, including the most recent one released in April 2016, the White Paper on Norway's Energy Policy.[12] A lot of efforts is put in research and development (R&D), but the lack of concrete long-term political goals for deployment of technologies is slowing down the level of activity in the industry.

11. NREAP, 2012.
12. *White Paper on Norway's Energy Policy: Power for Change*, Meld.St. 25 (2015–2016) (*Kraft til endring, Energipolitikken mot 2030*).

[C] The Role of Offshore Wind Power in Policy Papers

[1] Policy and Programs

A major step towards the development of offshore wind in Norway was the adoption of a dedicated regulatory framework in the form of the Offshore Renewable Energy Production Act in 2010. The Act is the result of an initiative from the coalition government in place at that time. It builds on an agreement reached by the coalition government during the parliamentary discussion of Report to the Storting No. 34 (2006–2007), *Norwegian Climate Policy* that a national strategy should be formulated for electricity production from offshore wind power and other marine renewable energy sources. An important element of that strategy was the adoption of a legislative framework necessary for granting licenses for offshore production facilities. This resulted in the adoption of the Offshore Renewable Energy Production Act of June 4, 2010 (hereafter OREP), also referred to as the Offshore Energy Act.[13] Those documents still remain the clearest and most ambitious policy documents adopted by the government on offshore wind.

The most recent document detailing the energy strategy of the government, the 2016 White Paper on Energy Policy,[14] refers only to a very limited extent to offshore wind power and does not contain any detailed target or vision. Similarly, the latest document detailing the climate change policy of the current government does not address offshore wind power in Norway.[15]

On the contrary, a large part of the research programs are dedicated to offshore wind. The steering committee of the program "Energi21" adopted a national strategy for research, development, demonstration and commercialization of new climate-friendly technologies which, in its revised version from 2014, recommends giving priority to, among others, offshore wind power.[16]

This policy priority was reiterated in the recommendation adopted by the Energy and Environment Committee of the Parliament in 2015, based on proposal of a Member of Parliament.[17] The Parliamentary committee requested the government to put forward a strategy contributing to the realization of demonstration projects for floating offshore wind and other forms of ocean-based renewable technology, as well as to investigate opportunities for the Norwegian supply industry within renewable energy generation.[18]

Industrial wind power actors in Norway are critical towards the lack of initiatives and vision of the government. The rejection of an offshore wind farm project at Siragrunnen (200 MW) and the absence of strategy on offshore wind power in the April

13. The Act entered into force on July 1, 2010.
14. *See* above 14, *White Paper on Energy Policy*, 2016.
15. Meld. St. 13 (2014–2015) *om ny utslippsforpliktelse for 2030 – en felles løsning med EU.*
16. *See* above 14, *White Paper on Norway's Energy Policy*, 2016, p. 61.
17. Representantforslag 118 S (2014–215) fra Stortingsrepresentant Rasmus Hansson, of May 15, 2015.
18. Innst. 70 S (2015–2016) *Innstilling til Stortinget fra energi- og miljøkomiteen*, Dokument 8:118 S (2014–2015).

2016 White Paper on Energy raised concerns as to the ambition level of the government. Most efforts have been concentrated on R&D,[19] but the domestic market is still lacking a breakthrough. The wind energy sector requires that efforts need to be put on demonstration projects as a first concrete step.[20]

In the aftermaths of the 2016 White Paper, the Parliament, who asked the government to come forward with a detailed policy on renewable energy at sea, noticed that the government did not follow-up its request and required the government again to come back with concrete proposals at the latest in 2017. The Energy and Environment Committee of the Parliament requested the government to elaborate a concrete strategy, including support measures, for the realization of demonstration projects for floating offshore wind and other forms of ocean-based renewable energy technologies.[21] As part of the 2017 state budget agreement, the Parliament requested the government to put forward a strategy for the commercial development of floating wind which can contribute to the profitable electrification of the offshore petroleum activities on the Norwegian continental shelf. Such strategy should be put forward at the latest in the state budget for 2018.[22]

[2] Quantitative Approach: Target-Setting for Offshore Wind

There is no overall quantitative target set for wind power in Norway. There is neither quantitative targets defined by the Ministry of Petroleum and Energy (MPE) in relation to Norway's energy agency, Enova, which is in charge of supporting innovation within energy and climate technologies.

Meanwhile, in the Strategic Impact Assessment completed in accordance with the 2010 OREP Act, the public authorities identified fifteen areas which can be opened to offshore wind power. The fifteen zones are estimated to have a capacity from 4600–12600 MW, with an estimated normal production of 19–50 TWh.[23] Due to technology improvements, this assessment can certainly be revised on the increase. Offshore turbines are increasing in size, with larger turbine blades and increased installed capacity, and those figures will probably increase.

[3] Qualitative Approach: Statements in Policy Papers

In the absence of quantitative target, Norway's policy has focused on a qualitative approach to offshore wind development.

---

19. NOK 750 million have been spent on research on offshore wind technologies since 2005 according to the Research Council of Norway.
20. Norwea, *Notat: Offshore 2025 – En bakgrunn for en norsk strategi for demonstrasjonsanlegg for havvindteknologi*, 2016.
21. Vedtak 869, Meld.St.25 (2015–2016), Innstilling (2015–2016).
22. Admodningsforslag fra Høyre, Fremskrittpartiet, KrF og Venstre, Point 10, day 2017.
23. SEA, NVE, 2013.

An alleged objective of the current government is to ensure the long-term development of *lucrative* wind power in Norway.[24]

The Norwegian government aims to develop a wind power strategy that minimize the conflicts between wind power and other interests.[25]

The government also sets itself as general goals for wind power to:[26]

- Ensure a better steering of the permitting procedures (licenses).
- Ensure a better control over the locations where project developer apply for concession.
- Ensure that the best locations are chosen, taking the quality of wind resources as a starting point and today's plans for grid infrastructures development.
- Ensure that environmental considerations are duly taken into account.
- But that a national framework for wind power is not a "Joint plan" for wind power.

Concerning offshore wind, the current position of the Norwegian government is that:

- Fifteen offshore areas have been identified by the Norwegian Water Resources and Energy Directorate (NVE).
- The costs are still higher than for onshore wind power.
- It is necessary to develop demonstration projects before being able to move from research to commercialization.

[4] Regional Cooperation, the Way Forward?

In June 2016, the Norwegian government signed a Political Declaration on energy cooperation between the North Seas Countries. The Declaration follows up previous initiatives such as the Memorandum of Understanding of December 3, 2010 on the North Seas Countries' Offshore Grid Initiative (NSCOGI), which resulted in the development of concepts around a possible offshore electricity grid in the North Sea. The June 2016 Declaration targets specifically offshore wind power as a potential to be exploited jointly through regional cooperation. It echoes the 2016 Manifesto "Northern Seas as the Power House of North Western Europe" signed by twenty Members of the European Parliament from countries neighboring the North Sea, and which called similarly for increased regional cooperation. In the June 2016 Declaration, the States set as an objective:

> to facilitate the further cost-effective deployment of offshore renewable energy, in particular wind, through voluntary cooperation, with the aim of ensuring a

24. According to the government, the national energy policy should enable "the long-term development of profitable wind power in Norway", which means that the market should come out with the best projects. *See* above 14, *White Paper on Norway's Energy Policy*, 2016, section 15.3.
25. *See* above 14, *White Paper on Norway's Energy Policy*, 2016, section 15.3.
26. *See* above 14, *White Paper on Norway's Energy Policy*, 2016, section 15.3.2.

> sustainable, secure and affordable energy supply in the North Seas countries, thereby also facilitating further interconnection between North Seas countries and – whilst focusing on a step-by-step approach – with the perspective of further integration and increased efficiency of wholesale electricity markets in the longer term, contributing to a reduction of greenhouse gas emissions and in average wholesale price spreads and to enhanced security of supply in the region.

§5.02 INSTITUTIONAL DESIGN FOR OFFSHORE WIND

[A] The "Main" Authority During the Development of Offshore Wind Power

The two main competent authorities are the Ministry of Petroleum and Energy (MPE) and the Norwegian water resources and energy directorate (Norges vassdrags- og energidirektorat, NVE).

MPE is responsible for both petroleum and energy policy. In the energy domain, MPE ensures the sound management, in both economic and environmental terms, of water and hydropower resources and other domestic energy sources. On behalf of the government, MPE acts as the owner of the Transmission System Operator (TSO) Statnett and Enova.

Established in 1921, NVE is a directorate and is responsible for the management of Norway's water and energy resources. NVE is the national independent regulatory authority for energy. NVE is a subordinate agency of the MPE responsible for the management of the energy and water resources on mainland Norway. The main statutory objectives for NVE concerning energy, and which the regulatory functions is a part of, is to promote social and economic development through efficient and environmentally sound energy production, and promote efficient and reliable transmission, distribution, trade and efficient use of energy. NVE heads contingency planning for power supply. It has duties regarding research and development and international cooperation. NVE holds the managing responsibility according to the Energy Act and the Water Resources Act. NVE assists MPE implementing the Industrial Licensing Act and the Act Relating to Regulations of Watercourses. NVE has the legislative power to issue regulations and to make individual decisions and perform preparatory procedures of cases to be resolved by the MPE. NVE has delegated powers according to the Energy Act.

The following ministries are also consulted:

- Ministry of Climate and Environment is responsible for environmental policy and has an overall responsibility for climate policy.
- Ministry of Industry and Commerce.
- Ministry of Fisheries and Coastal Affairs.

[B] The "Supplementary" Authorities During the Development of Offshore Wind Power

[1] Central Government Level

During the phase of preparation of the 2010 study proposing an SEA-assessment program for offshore wind energy outside the baseline, an inter-directorate group chaired by NVE was established. It also consisted of the Norwegian Directorate for Nature Management (now Norwegian Environment Agency), the Norwegian Directorate for Fisheries, the Norwegian Coastal Administration and the Norwegian Petroleum Directorate. The inter-directorate group reported to a steering committee led by the MPE, which also consisted of the Ministry of the Environment and the Ministry of Fisheries and Coastal Affairs. The same directorates and government bodies, including NVE, contributed actively during the phase of elaboration of the SEA.

[2] The Role of the Local Government

Local governments, and primarily municipalities, have a role to play if the offshore wind installations are to be installed within the baseline. They are central actors in the decision-making process under the Energy Act and the Planning and Plan and Building Act (planning and building permits), which is decisive for the approval of the project. Another important aspect is that they can require compensation measures for the projects.

County councils can adopt county plans for the location of wind power projects and have been encouraged to do so by the government. County plans serve as guidance for the elaboration of sectoral and municipal plans. The objective is to better anticipate conflicts in area use when planning the location of wind energy installations.[27]

[3] Other Players Vital for the Development of Offshore Wind

[a] State Organs

Enova SF is a public enterprise, the task of which is to promote energy efficiency, energy savings, new renewable sources of energy and environmental-friendly use of natural gas. MPE acts as its owner on behalf of the government. MPE also defines Enova's tasks and goals in an annual mission letter.

27. See H.-C. Bugge "Norwegian Planning Law", in H. Tegner Anker, B. Egelund Olsen and A. Rønne, *Legal Systems and Wind Energy: A Comparative Perspective* (Kluwer, 2009), p. 133.

[b] *Industry Association*

Norwea, Norwegian Wind Energy Association

Energy Norge (Energy Norway) is a non-profit industry organization representing about 270 companies involved in the production, distribution and trading of electricity in Norway.

Norsk Industri is the Federation of Norwegian industries.

Norges Rederiforbund (Norwegian Shipowners' Association) is a trade and employment organization for Norwegian controlled companies within the shipping and offshore industry.

Norwegian Energy Partners (NORWEP) is the result of a merger between INTSOK and Intpow in 2017, with the purpose of defending the Norwegian energy industry with a focus on overseas markets.

Wind Cluster Norway is an industry *organization* which aims to improve profitability and increase value creation for the cluster companies by developing solutions for reduced cost of energy in the value chain for wind power production. The project aims to strengthen Windcluster Norway's position as an internationally recognized cluster within production of wind power in harsh environments.

[c] *Research Centers*

There are two main research centers for offshore wind in Norway, which are NOWITECH and NORCOWE. CEDREN is also conducting related activities. All three centers are funded half by the Research Council of Norway and half by the industry and the research institutions.

The Research Center for Offshore Wind Technology (NOWITECH) is hosted by SINTEF Energy Research. The objective of NOWITECH is pre-competitive research laying a foundation for industrial value creation and cost-effective offshore wind farms. Emphasis is on "deep-sea" (+ 30 m) including bottom-fixed and floating wind turbines. NOWITECH is part of the Centre for Environment-friendly Energy Research (FME) scheme cofunded by the Research Council of Norway. The FMEs are established to conduct concentrated, focused and long-term research of high international caliber in order to solve specific challenges in the field of energy and the environment.

The Norwegian Center for Offshore Wind Energy (NORCOWE) is led by Christian Michelsen Research.

Although not focusing solely on offshore wind, the Center for Environmental Design of Renewable Energy (*CEDREN*) also conducts research on environmental issues within wind energy and other renewable energy production.

[d] Commercial Actors

The Norwegian industry takes part in component production for wind energy systems (e.g., wind turbine blades and nacelles) on a very limited scale.[28] Companies with experience from the offshore oil industry, such as Kvaerner, OWEC Tower and Aker Solutions, have developed several projects within the foundation market. These companies offer offshore wind turbine substructure solutions like jacket quatropods and tripods, and foundation design. Increased construction of wind farms will generate engineering and construction jobs, and ultimately jobs for maintenance personnel. Some Norwegian energy companies are already present in the offshore wind market abroad, as operator or owner, in the perspective of gaining experience (primarily in the United Kingdom). The key companies, both suppliers, owners and operators, are listed below.

Statoil has developed an ambitious strategy within offshore wind. Statoil, with its experience in offshore oil and gas production, has developed the world's first full-scale floating turbine, Hywind, and is involved in the operatorship of several wind farm parks in the United Kingdom.[29] Statoil is 67% owned by the Norwegian state. The ownership interest is managed by the MPE.

Statkraft is Norway's largest and the Nordic region's third largest power producer. It is wholly owned by the Norwegian state. The company has recently reassessed its strategy within the offshore wind market. In March 2017, Statkraft decided to exit the Dogger Bank offshore wind projects. It sold its 25% interest in the 4.8GW Dogger Bank offshore wind projects to partners Statoil and SSE. The sale of the Dogger Bank projects is in line with Statkraft's strategy to explore opportunities to exit the ownership of offshore wind power assets. However, Statkraft still believes that the UK offshore wind sector presents a significant opportunity to develop a secure, sustainable, cost-competitive energy source and announced that it will continue to participate as an enabler of the development and operation of projects through entering into power purchase agreements. Stakraft still holds shares in other UK projects: a 40% share in Sheringham Shoal Offshore Wind Farm, 30% in Dudgeon Offshore Wind Farm and 50% in the Triton Knoll Offshore Wind project.[30]

Kvaerner (Verdal construction site) has supplied components to jacket and tripod foundations. The jacket segment is expected to grow in the new future.

OWEC Tower is a company that designs jacket foundations and has completed both commercial and demonstration projects.

Fred. Olsen Ocean and *Fred. Olsen Wind Carrier* have been active in the turbine installation segment, making use of their own jack-up vessels.

The segment of cable manufacturing is very active. Although not having Norwegian mother companies, the Norwegian subsidiaries of Nexans, Parker Scanrope and Draka (Prysmian) have manufactured a large amount (almost one-third) of the cables

28. *See* the market overview conducted by MAKE Consulting for INTPOW, Export Credit Norway and Greater Stavanger, *Norwegian Opportunities in Offshore Wind*, 2017.
29. *See* above section B.
30. Statkraft, press release, 23.03.2017.

for offshore wind farms since 2010. Nexans is particularly strong on export cable manufacturing.

Norwegian cable installation companies like DeepOcean and Siem Offshore have good track records. Siem Offshore is also a cable vessel supplier, which is another promising market for Norway.

Aibel has entered the HVDC substation segment and was contracted to construct the HVDC substation, Dolwin Beta (while subcontracting the construction to Drydocks World in Dubai).

Sway Turbine, another company involved in the offshore wind power, has gone bankrupt.

[e] System Operators

Statnett SF is the transmission system operator (TSO). It owns about 87% of the transmission grid. It is owned by the state and supervised by MPE. Statnett's revenues are regulated by NVE, as part of NVE's regulation of monopoly operations. Concerning installations located outside the baseline, the OREP Act contains provisions as to system operation. The Ministry can nominate a system operator who will be responsible for ensuring balancing operations. However, Statnett has not been particularly active on the topic yet.

Distribution System Operators (DSOs) – There are 146 network companies that own and operate regional distribution networks. Some also own minor parts of the transmission network.

[f] Investors

Several organisations are active in promoting the export activities of Norwegian companies, including within the offshore wind sector: Innovation Norway, GIEK, Eksportkreditt Norge. Some European investment funds and banks have started investing in onshore wind energy projects in Norway, but none within the offshore wind market.

§5.03 LEGAL DESIGN FOR OFFSHORE WIND

[A] Incentives

[1] Main Finance Scheme

Norway has no support regime specific to offshore wind at the moment. Offshore wind power projects can therefore apply to the green certificates scheme or R&D / technology development support, as any other renewable energy project.

[a] Change of Support Regime

Norway introduced a new support scheme in 2012. Before that, and since 2001, Enova SF has been in charge of providing financial support to renewable energy projects, including wind power projects, on a case-by-case basis, with the goal to support projects just enough to make them commercially viable.

[b] The Green Certificates Scheme

Norway adopted a new support scheme in 2011, which came into force on January 1, 2012.[31] It is a green certificate scheme, based on a joint green certificates market with Sweden. Under the Norwegian scheme, electricity generators using eligible renewable technology receive one certificate (called elcertificates) for each MWh of electricity generated. The demand is created by a legal obligation for electricity suppliers and big electricity customers to surrender elcertificates equivalent to a predetermined proportion of the total electricity they supply or use (quota obligation). If the obliged suppliers do not possess enough elcertificates (e.g., through generation from their own eligible plants), they can purchase the necessary certificates from eligible generators, either bilaterally or on the market (traders or organized markets). The price of certificates is determined in the market by supply and demand, and varies from one transaction to another. The level of the quota obligation is adjusted every year and is consistent with a curve determined by the energy regulator NVE in order to meet target achievement. Although the scheme remains national, the elcertificates can be traded on the joint Norwegian-Swedish elcertificates market (also operative since January 1, 2012). The functioning principles of the joint market are set in a Bilateral Treaty.[32] The economic incentive is designed to stimulate the combined development of 26.4 TWh/yr of new renewable power production in the countries together. The burden is split equally, which means that 13.2 TWh will be developed in each of the two countries by 2020.

The scheme is technology neutral, meaning that all technologies - including offshore wind - are eligible under this scheme. The only restrictive criterion for access to this scheme is that the renewable energy plant must have started construction after September 7, 2009. Hydropower plants with an installed capacity below 1 MW and whose construction started after January 1, 2004 are eligible for elcertificates. Wind, water and bio energy are the most important energy resources in the certificate market, but has not developed equally in the two countries (more wind and bioenergy in Sweden and more small hydropower in Norway). Offshore wind has not benefited from this scheme at all, because it is not economically attractive. Norway is committed to finance 50% of the certificates, regardless of where the production is located in each of

31. Law on Electricity certificates (*Lov om Elsertifikater*) and Regulation on Electricity certificates (*Forskrift om elsertifikater*).
32. Agreement between the Government of the Kingdom of Norway and the Government of the Kingdom of Sweden on a common market for electricity certificates, June 29, 2011.

the two countries.[33] The costs of financing the scheme are borne by final customers, in the electricity bill.

To avoid over-subsidies, eligible power plants will be entitled to earn certificates for fifteen years, starting from their commissioning date. The scheme will close in 2020, and the last electificates will consequently be issued in 2034, the scheme closing on December 31, 2035. Based on a political consensus across political parties, it has been decided not to renew the scheme after that date.[34]

[2] Supplementary Finance Scheme

Other sources of financing include: Enova, Innovation Norway, Norwegian Council of Research, Export Credit Norway and GIEK.

[a] Low Interest Loan and Loan Guarantee

Eksportkreditt (Export Credit Norway) offers Norwegian and foreign companies competitive financing when buying goods and services from Norwegian exporters. Export Credit Norway is wholly owned by the Norwegian Government and is administered by the Ministry of Trade, Industry and Fisheries. Export Credit Norway offers loans with two different sets of interest terms: fixed-rate Commercial Interest Reference Rate (CIRR) loans and CIRR-qualified market loans with variable rates. Both types of loans may be disbursed in all OECD currencies, and both comply with the OECD Arrangement on Officially Supported Export Credits.

The Norwegian Export Credit Guarantee Agency (GIEK) is a public enterprise that reports to the Ministry of Trade, Industry and Fisheries. It is run as a financial enterprise that issues guarantees on the same conditions as banks. The Norwegian state stands behind GIEK's guarantees. GIEK's mandate is to promote Norwegian exports and investments by providing long-term guarantees on behalf of the Norwegian state.

[b] Tax Credit

There is no tax credit regime specific to offshore wind. Paradoxically, the petroleum taxation regime could be an attractive manner for offshore wind project to develop. The Siragrunnen project intended to make use of that taxation regime by supplying electricity to the offshore petroleum platforms. As of today, it is a case-by-case analysis, subject to the exact plan for the project (Plan for Development and Operation).

---

33. NREAP, 2012.
34. The situation is different in Sweden where, in June 2017, the parliament decided a new target for electricity from renewable energy sources until 2030. The system shall consequently increase the production of electricity from renewable energy sources by further 18 TWh by 2030.

[c] *Investment Subsidy/Grant*

The Norwegian energy agency, *Enova*, offers capital grants for full-scale demonstration projects for ocean renewable energy production including offshore wind. While up to 50% of eligible costs can be covered, Enova's funding measured in absolute figures is limited. Since 2012, Enova has been focusing on supporting technology development connected to wind power.

Innovation Norway runs a program supporting prototypes within environmental friendly technology, and wind energy is included in this definition. Projects are supported with up to 45% of eligible costs.

[d] *Favorable Grid Cost-Sharing Rules*

The same rules for any other wind power project apply to offshore wind, but are neither necessary nor beneficial.

[e] *The Links to ETS, CDM and JI*

At the moment, there is no electricity generation from offshore wind farms in Norway. The question has consequently not been raised.

[f] *Other Funding Source, Such as National Research Fund and Supportive Scheme*

Paradoxically, offshore wind power represents a large part of the public and private research budget.

The *Research Council of Norway* administers a public research program for sustainable energy, ENERGIX. This program covers renewable energy, energy efficiency, energy systems and sustainable transport (hydrogen, fuel cells, biofuels and batteries). Industry, research institutes and universities may receive funding for their research through proposals to regular calls.

[g] *International Trade Law Concerns of the Subsidy*

There are no particular international trade law issues arising from the design of the subsidy scheme in Norway. There is not any apparent local content requirements in Norwegian legislation either.

[3] *Incentive for Construction Harbor, Construction Vessel and Grid, or Turbine Manufacturers*

Norges Rederiforbund, the shipowner association, is currently working on a review of the taxation regime for vessels. This includes a review of the taxation regime for carriers for offshore wind platforms.

[B] Regulations

Two Parallel Regimes According to the Location of the Installations

The regulation of offshore wind in Norway is subject to two different regimes according to the distance of the installation from the shore.

If the offshore wind farm is to be installed within the baseline, it falls within the scope of application of the Energy Act of June 29, 1990 (section 1-1, Energy Act) which applies to all forms of electricity production.

If the project is to be developed outside the baseline and on the Norwegian continental shelf, it falls under the scope of application of the OREP Act of June 4, 2010 (section 1-2, 2010 OREP Act). The OREP Act applies in the Norwegian territorial waters outside the baselines and in the exclusive economic zone (EEZ), but certain provisions of the Act can also be applied in the coastal waters inside the baselines - the inner waters (section 1-2, paragraph 7).[35] The Act also applies to all forms of renewable energy production at sea. The 2010 OREP Act regulates the production of offshore renewable energy, the transformation and the transmission of the electricity produced offshore, including the installations related to those operations (section 1-1).

[1] *License Scheme*

[a] *Within the Baseline: Energy Act*

For all new energy projects, including wind power and related infrastructures, a license to build and operate is required. The procedures differ according to energy source and size, but the general rule is that energy projects located within the baseline require a construction and operation license under the Energy Act.

Therefore, if offshore wind turbines are to be installed within the baseline, a license is required pursuant to the Energy Act. The developer will be responsible for putting forward the application. The relationship between owner and developer, subject to private arrangement, is not regulated and must only be referred to in the license application. The Act defines minimum competence criteria for becoming a licensee.

35. Note that Norway established an EEZ by passing Act of December 17, 1976 relating to the Norwegian exclusive economic zone.

As a derogation to this general rule, a wind park up to five turbines and of total installed effect under 1 MW will not be subject to licensing rules. Simpler requirements will apply. Those projects will be dealt with by the municipalities directly under construction/building legislation (Building and Construction Act 2009).

For offshore wind projects of more than five turbines, of total installed effect above 1 MW and located within the baseline, the general licensing procedure applicable under Energy Act will consist in five main steps:

- *Step 1 – Notification and Environmental Impact Assessment (EIA) program.* The project developer will notify the planned project to the licensing authority, i.e., NVE and third parties possibly affected by the project. In the notification, the project developer will describe the project. Third parties are invited to comment on the planned project. The notification contains a proposal for EIA program, identifying the issues to be subject to the assessment.
- *Step 2 – Public hearing of notification and approval of the EIA program.* On the background of the comments received, NVE will determine the EIA program.
- *Step 3 – License application, EIA and public hearing of license application and EIA.* After completion of the EIA, the project developer can elaborate its license application. The application contains a description of the project and the results from the EIA. NVE submit the received application and the EIA again to public hearing to the relevant parties.
- *Step 4 – License decision.* NVE reaches a decision, on the basis of all received documents, the results from the public hearing and its own assessment.
- *Step 5 – Possibility of appeal.* The license decision can be appealed to the MPE.

The whole procedure usually takes two to three years, to which a possible appeal procedure to the MPE must be added.

So far, only one commercial licensing for offshore wind project has been awarded under the Energy Act following the abovementioned procedure (Havsul I project, developed by Vestavind Offshore, awarded in June 2008). However, the investors in the project - mostly local municipalities - decided to cancel it because of lack of financial certainty. NVE has dealt with other license applications - Siragrunnen, Havsul II, Havsul IV - but those projects have not been awarded a license. Many other offshore wind projects have been discussed, but have not been processed and have stopped before license application, at the notification stage (step 1): Utsira (Lyse Produksjon AS), Ægir (Oceanwind AS), Mørevind (Trønderenergi Kraft AS), Steinshamn (Offshore Vindenergi AS), Sørlige Nordsjøen (Lyse Produksjon AS), Selvær (Nord-Norsk Vindkraft AS), Lofoten havkraftverk (Lofotkraft Vind AS), Idunn (Fred Olsen Renewables AS), Stadtvind (Vestavind Kraft AS), Gimsøy (Lofotkraft Vind AS), Vannøya havkraftverk (Troms Kraft Produksjon AS), Havsul III (Havgul AS).[36] In some of those cases, NVE asked the project to be stopped because of lack of grid capacity (e.g., Fosen) or requested the applicant to amend the project. In a series of cases dated 2008 before the adoption of the OREP Act, NVE was enabled to deal with the

36. NVE website, list of license cases <https://www.nve.no/konsesjonssaker/>.

applications because the projects were located outside the baseline (Ægir, Mørevind, Idunn, Sørlige Nordsjøen, Stadtvind).

NVE has awarded licenses for three demonstration project within the baseline to Marin Energi Testsenter AS for a period of twenty years from the date of entry into function (located outside Kvitsøy, Rennesøy og Karmøy).

[b] Outside the Baseline: OREP Act

Pursuant to the 2010 OREP Act, the right to exploit offshore renewable energy resources belongs to the Norwegian State (2010 OREP Act, §§1–3). This Act is based on public administration and control of the management of the energy resources offshore. It is based on the principle that the right to exploit renewable marine energy resources belongs to the Norwegian Government.

Opening of Areas

Before awarding any license, the licensing areas must be formally opened to the exploitation of renewable energy resources after the completion of an environmental impact assessment (section 2-2 OREP Act) (for details on EIA, *see* section B[2][b] below). The decision to open the areas for exploitation officially pertains to the King, by decision of Council of State (section 2-2, 2010 OREP Act). The proposal for opening of licensing areas is made public and subject to public consultation. Exceptions may, however, be made from the rules on opening of areas in special cases.

Licensing Procedure

Outside the baseline, the establishment of installations for the production, transformation or transmission of electricity in the designated areas is made dependent on a license (Chapter 3).

The licenses may stipulate terms and condition for energy supply, environmental concerns, safety, commercial activity and other interests in connection with the development, operation and decommissioning of such facilities. Conditions may be stipulated for facilitating connection to other installations or systems. It may also be stipulated that the licensee must furnish a guarantee for the fulfillment of the obligations it undertakes.

A concession is needed to the building or extension of production facilities (section 3-1) and the transmission facilities (section 3-2). Once the area is opened and the concession has been awarded, a *Detailed Plan* must be submitted to the Ministry for approval, and for both production facilities and transmission facilities. The Detailed Plan is itself subject to an environmental impact assessment (*see* section below). The Detailed Plan shall contain an account of the technical, safety and environmental aspects of the installation.

The licensee must be the legal person/entity established under Norwegian law and registered in the Register of Business Enterprises (section 3-5), if not provided otherwise by international agreements.

The establishment of installations in these areas can be made subject to an area fee (Chapter 10).

Similar to the situation for offshore petroleum installation, the Act contains provisions governing offshore safety, working environment and the establishment of safety zones as well as the abandonment and removal of OREP installations and offshore networks (Chapters 5 and 6).

Duration

For both production facilities and transmission facilities, the license is awarded for a period of thirty years from the date the installation enters into function. The licensing period may be extended, based on an application of the license holder.

Decommissioning of Installations

Pursuant to Section 6-1 OREP Act, when the installation is closed down, all parts of the installation shall be removed, except not decided otherwise by the MPE. The licensee for the electricity production facility shall, in good time before the end of the concession, submit a decommissioning plan to the MPE.

Concession for the Import and Export of Electricity

Specific rules regarding import and export of electricity produced offshore are included in Chapter 8, OREP Act. The import or export of electricity (from for example offshore wind facilities) to or from a foreign State requires the award of a license (section 8, OREP Act).

Practice

MPE has awarded two concessions for demonstration projects for floating offshore wind: Hywind, developed by Statoil ASA and move moved from inside to outside the baseline, close to oil and gas platforms; and another project outside Karmøy developed by Marin Energi Testsenter AS.

[c] Assessment of the Regime

Within the baseline, the recent Siragrunnen project can serve as a good example of the remaining challenges faced by offshore wind project developers. The project consisted of sixty-seven wind turbines for a total of 790 GWh generation per year. The project also included the establishment of an online manufacturing plant for wind farm fundaments. This would have contributed to solve of the issue of lacking full-scale demonstration project for offshore wind in Norway. NVE rejected the license application in January 2015. One main reason for NVE to refuse awarding the license was the costs level (over NOK 6 billion), which the NVE deemed not to be socioeconomically rational (one of the criteria in the licensing procedure). Other elements were the impacts on biodiversity and fisheries. The decision was appealed to the MPE, which confirmed NVEs decision in 2016. There is consequently an interest for offshore wind projects within the baseline, but conflicts in area usage and costs levels remain major hinders.

Outside the baseline, the adoption of the OREP Act is now providing for a clear and transparent regulatory regime, but the costs for developing offshore wind projects in such deep waters do not make it economically attractive yet for investors.

[2] Environmental Impact Assessment (EIA)

[a] Within the Baseline

If the offshore wind power installation is located within the baseline, there is a threshold of 1 MW Energy Act for completing a full EIA. The procedure for completing the EIA has been described above (*see* section B[1][a]). To facilitate the elaboration and implementation of the EIA-programme, NVE has released several guidelines, in particular the Guidelines for planning and situating wind power installations.[37] The latter Guidelines apply to all wind power projects, but do not deal with the specificities of offshore wind, subject to case-by-case analysis (e.g., conflicts with shipping, fisheries, aquaculture).

[b] Outside the Baseline

If the offshore wind power installation is to be built outside the baseline, it will always be subject to an impact assessment.

After the area has been opened and the concession has been awarded, a Detailed Plan must be submitted to the Ministry for approval, for both production facilities (section 3-1, OREP Act) and transmission facilities (section 3-2, OREP Act). The Detailed Plan is itself subject to an impact assessment, which must be annexed to the application for a license. Both the Detailed Plan and the EIA are subject to the formal approval for the MPE.

The OREP Act also contains provisions for impact statements in connection with license applications and detailed plans. Before the work on the impact assessment starts, a proposal for assessment program must be submitted to the public before the work can start. The Impact Assessment is not limited to environmental aspects, but encompasses also economic, commercial, cultural, community interests, among others.

If a decision adopted in accordance with the OREP Act can have negative environmental consequences for another State, the MPE has the obligation to inform that State which then has the opportunity to influence the plan or impact assessment work (section 4-2, OREP Act).

---

37. Ministry of Petroleum and Energy and Norwegian Environment Agency, *Retningslinjer for planlegging og lokalisering av vindkraftverk* (T-1458, 2007).

[3] SEA and Siting Related Planning

[a] Within the Baseline

Norway has not adopted a national plan (i.e., strategic environmental assessment) for wind power onshore yet. Due to the need to better coordinate the license applications and the national priorities, MPE has given NVE the mandate to come forward with a proposal for national plan to be delivered by December 2018 after a phase of public consultation.[38] The plan will focus on onshore wind, and it is still unclear to which extend offshore wind within the baseline will be included.

Although there is no national SEA as such, county councils may elaborate regional plans to ensure a comprehensive and long-term development for wind power in their region. Regional plans should provide guidelines for appropriate planning and site selection of wind power plants within the county or region. The regional plans focus on identifying and characterizing the potential for conflicts, and should strengthen the foundation for a comprehensive assessment of wind-power projects for licensing. MPE has elaborated short guidelines for the elaboration of the regional plans and included them in the Guidelines for planning and situating wind power installations.[39] The existing regional plans have been assessed as complying with the main requirements of the SEA Directive 2001/42/EC by external consultants.[40]

[b] Outside the Baseline

The development of offshore wind energy outside the baseline is restricted to designated areas. Pursuant to the OREP Act, the construction of offshore wind power (as other renewable energy production units/facilities at sea) can only take place after the Norwegian government has opened specific geographical zones for license applications. The opening of zones requires that an SEA is carried out. The SEA is carried out on an overall level and is not a substitute for project-specific impact assessments. The goal of the SEA is to provide the best possible basis for deciding which of the zones should be opened for license applications. The process for identifying those suitable areas is further described in section 2-2 of the OREP Act. The areas are selected based on technical and economic criteria. The areas must also give rise to limited conflict of interests.

Shortly after the adoption of the OREP Act, the process for elaborating the first Strategic Environmental Impact Assessment for offshore wind started and was conducted in two phases by the NVE (2010 and 2013). The final SEA document was

38. This order is a direct follow-up of the White Paper on Norway's Energy Policy (Meld. St. 25 (2015-2016)). See the order letter by MPE: *Bestilling - nasjonal ramme for vindkraft*, 9.2.2017.
39. Ministry of Petroleum and Energy and Norwegian Environment Agency, *Retningslinjer for planlegging og lokalisering av vindkraftverk* (T-1458, 2007).
40. R. May, *Strategisk konsekvensutredning for landbasert vindkraft - En evaluering av regionale planer for vindkraft*, NINA, Report 746, 2011.

published in 2013. The following paragraphs retrace the main elements which have been discussed during the SEA process.

In 2010, after adoption of the OREP Act, NVE conducted a zone assessment work. Shortly after, in October 2010, NVE published a study on proposed areas for SEA. In its 2010 Assessment report, NVE proposed areas that may be suitable for the establishment of offshore wind power, and which ought to be considered more closely in a SEA, as envisaged in Proposition No. 107 (2008–2009) to the Odelsting. Among the areas identified, eleven are suitable for fixed-bottom installations and four are suitable for floating turbines. Among the factors taken into account were: engineering, environmental and user considerations. During its work, the inter-directorate group that was set up for the purpose of the 2010 study proposed a program for a SEA that covered the following topics: power production, power system and market; relationship to legislation, plans and conservation areas; natural environment, including seabirds, fish, marine mammals and seabed ecology; commercial and social interests, including fisheries and aquaculture, petroleum, shipping, cultural milieu, landscape, outdoor recreation, tourism, the Armed Forces, industry and employment and other use of areas; risk; total impact; and effects on other countries.

In January 2013, NVE presented to the MPE its study on SEA for Offshore Wind (*Havvind strategisk konsekvensutredning*). It is now up to the Ministry to decide which zones will be opened for license applications. The Assessment identified fifteen priority areas for developing offshore wind energy.

The document presented by NVE summarized the main findings of the SEA as well as NVE's recommendations to the OED. In particular, NVE decided, based on the results of the SEA, to define three categories of zones:

- **Category A**: Wind power development within the zone is technically and economically feasible and will have relatively few negative impacts. Grid connection is possible before 2025.
- **Category B:** Wind power development within the zones will have challenges related to either technical aspects or conflict of interests/negative impacts. The challenges might be resolved in the future through technology development, grid measures and/or mitigation measures. NVE considers that zones in this category can be opened when technology matures, or when existing use of the areas changes.
- **Category C:** Wind power development within the zone represents greater challenges than in the other two categories. Conflicts of interest in the areas are not easily resolved. Foreseen negative impacts are still considered acceptable. Zones in this category should not be opened at the expense of zones in the two other categories.

The existence or feasibility of power transmission system was also an important criterion when choosing the areas. The 2010 Study from the inter-disciplinary group had already defined three area categories based on that criterion, reflecting the access or need for adaptation to the power system:

(1) Wind-power areas connected with large reservoir power stations.
(2) Wind power areas that enable direct export to Northern Europe.
(3) Wind power areas that allow the safeguarding of regional (national) energy balances.

A potential area for offshore wind power ought to fall into at least one of the three categories above.

The result presented in the 2013 final report is a short-list of fifteen zones of a capacity from 4,600–12,600 MW, with an estimated normal production of 19–50 TWh. The areas differ in size. The two southern most zones, Sørlige Nordsjø I and Sørlige Nordsjø II, cover areas of 1,300 km^2 and 2,500 km^2 respectively. The areas of the zones considered for floating turbines range from 500–1,000 km^2, and the zones considered for bottom-fixed installations range from 50–300 km^2. In total, the areas considered cover a total area of 9,000 km^2, approximately 1% of the Norwegian EEZ.

Figure 5.1 Zones Considered for Offshore Wind Power in Norway

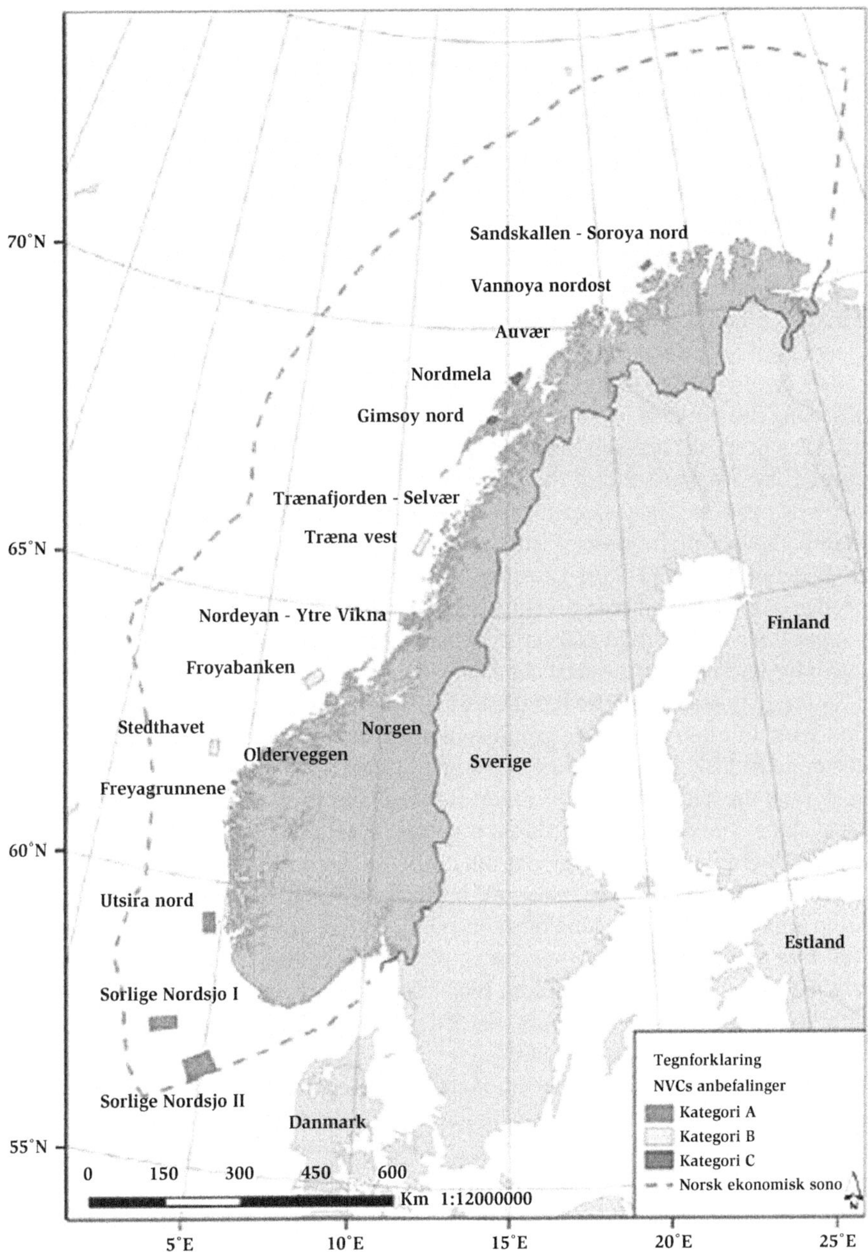

These Categories indicates NVE's recommendations for the further process based on the degree of possibilities and challenges regarding wind power development.

[4] *Planning the Legal Regime: Marine Planning and/or Land Planning*

[a] *Inside the Baseline*

Coastal waters within the baseline around mainland Norway cover an area of approximately 90,000 km^2. Meanwhile, only 2,900 km^2 of it is protected. During the past five to ten years, under both national, EU and international commitments, Norway has taken initiatives to strengthen its national legislation towards a more integrated marine protection and planning regime.

The main tool within the baseline is the adoption of the national Marine Protection Plan, which has as purpose to protect a representative sample of the sea areas along the coast, and to secure the diversity of species and biotopes. The work on the marine protection plan is a follow-up of Report No. 43 (1998–1999) to the Storting Conservation and use of the coastal zone (the Coast Report). It started in 2009 and has resulted in the development of joint criteria and the adoption of marine protection areas along the Norwegian coast to be included in the national plan (six as of February 2017). The proposed areas in the plan are meant to cover the full variety of Norwegian nature. They are divided into the categories of coves, rich current localities, special shallow-water areas, fjords, open coastal areas and areas reaching from the coast and out into the ocean (transects) and shelf areas. This protection regime has been introduced in the 2009 Nature Diversity Act,[41] and only applies to marine areas. If the objective is to protect an area covering both land and sea, other protection regimes will be applicable, such as national parks or natural reserves. The drafting and management of the national marine protection plan is a joint effort. The Norwegian Directorate for Nature Management (DN) has the responsibility for the plan centrally, in collaboration with an inter-directorate group consisting of representatives of the Norwegian Directorate of Fisheries, NVE, the Norwegian Coastal Administration, the Petroleum Directorate, the Directorate for Cultural Heritage and the Norwegian Defence Estates Agency plus the county councils represented by Hordaland County Council. The County Governors and the regional offices of the Directorate of Fisheries have the responsibility locally and regionally. According to §39 of the Nature Diversity Act, in a marine protected area no person must do anything that reduces the conservation value of the area as described in the purpose of protection. A marine protected area may be protected from all activity, pollution, projects and use, subject to the limitations that follow from international law. Any restrictions imposed on activity shall be proportional to the purpose of protection.

At local level, the management of coastal waters inside the baseline falls under the competence of the authorities for each river basin district. This follows from the Water Management Regulations which transposes into Norwegian law as the EU Water

41. §39, Nature Diversity Act of June 19, 2009 No. 100 relating to the Management of Biological, Geological and Landscape Diversity (Nature Diversity Act).

Framework Directive. The marine and coastal management regimes aim to better take into account the interactions between marine and coastal waters and their ecosystems. This results in the adoption of local marine management plans and river basin management plans which are tools for ecosystem-based management of marine and coastal waters and rivers and lakes.

Figure 5.2 Marine Protected Areas

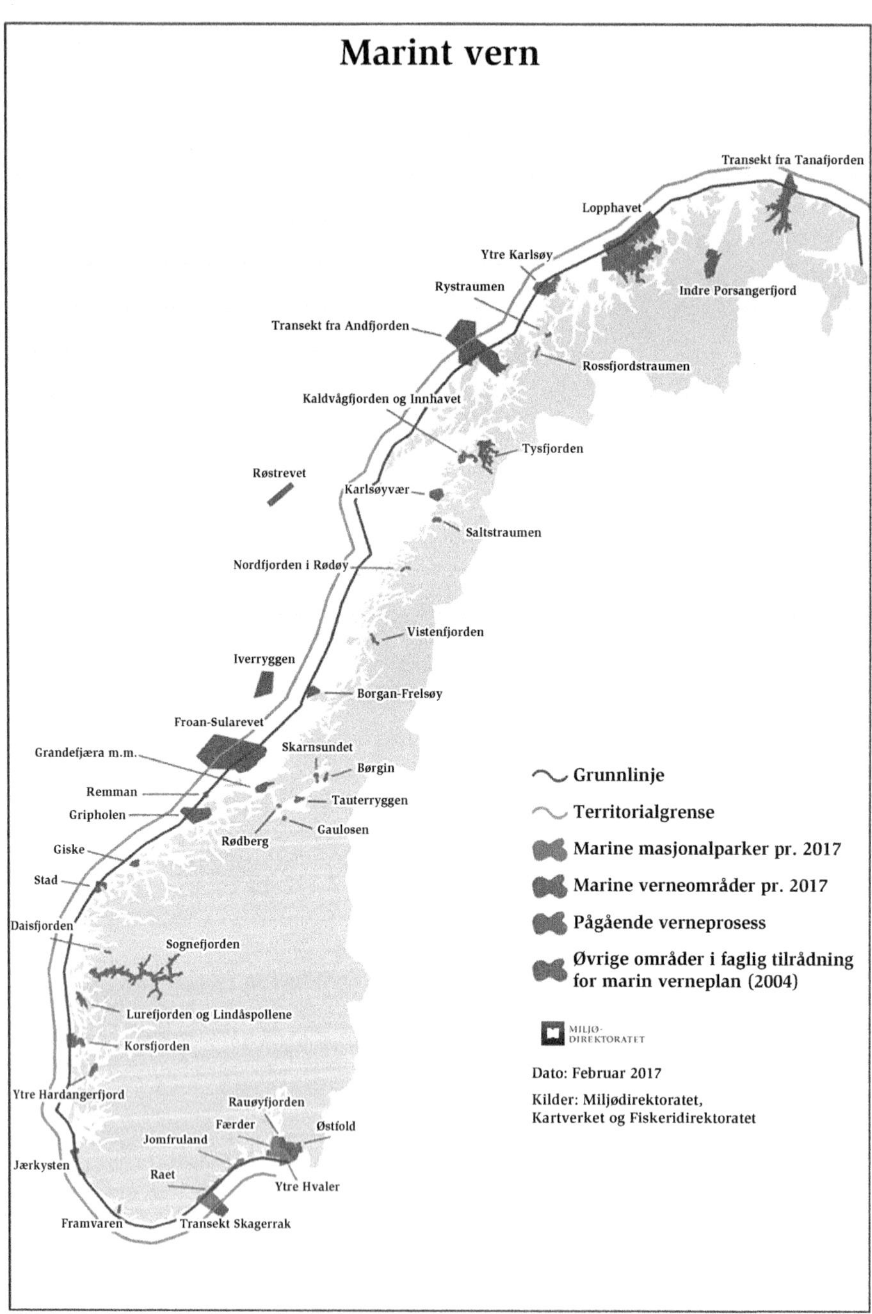

Source: The Norwegian Environment Agency.

[b] Outside the Baseline

During the last fifteen years,[42] the Norwegian government has strengthened its efforts to elaborate a more consistent strategy for the management of the marine areas under its jurisdiction.[43] This move corresponds to a general trend in terms of environmental management, which consists in applying an ecosystem-based management approach to the marine environment. It is also in line with Norway's international and EEA commitments, and notably under public international law: the United Nations Convention on the Law of the Sea (UNCLOS Convention) obligation to protect the environment while exploiting sea resources; the Convention on Biological Diversity; the Johannesburg Plan of Implementation (§30); and the UN Sustainable Development Goal 14.

Those efforts have resulted in the adoption of management plans for the following three marine areas: the North Sea and Skagerrak;[44] the Norwegian Sea;[45] and the Barents Sea including Lofoten.[46]

The purpose of the Norwegian Management Plans for marine areas is to facilitate value creation, the coexistence between industries and the sustainable harvesting of resources.[47] The plans contribute to the implementation of an integrated ecosystem-based management of the marine environment in the Norwegian waters. Ecosystem-based management is a well-known environmental management approach which has found echoes in public international law and which has been applied to the marine environment since the years 2000.[48] It entails that the management of human activities must take as a starting point the limits set by the ecosystem itself, so as to maintain its essential structure, functioning, production and biodiversity. It looks at the whole range of interactions within an ecosystem. In the Norwegian context, it shall facilitate the coexistence of different industries such as fisheries, shipping and petroleum operations within the relevant marine environment. The Management Plans cover waters from the baseline to the open sea as well as the human activities in those areas.

The elaboration of the Management Plans starts with ecosystem-based assessments for each of the main economic activities concerned in the area as well as an assessment of the interactions between the relevant commercial activities such as petroleum, fisheries and shipping. The plans also define measures to reduce the environmental burden of those activities or to solve competitive uses of the same sea

42. *See* in particular, White Paper nr. (2001–2002) *Rent og rikt hav*. Innst. S. nr. 161 (2002–2003) White Paper nr.19 (2004–2005) *Marin næringsutvikling. Den blå åker* (Innst. S. nr. 192 (2004–2005)).
43. For an analysis of the Management Plans for Norwegian Sea Areas, *see* H.C. Bugge, "Har vi de rettslige redskapene som trengs for en god forvaltning av våre havområder?", in M. Stub and I. Hjort Kraby (eds.), *Forsker og formidler. Festskrift til Erik Magnus Boe på 70-årsdagen 17. april 2013* (Universitetsforlaget, 2013), pp. 65–87.
44. Adopted in 2013 (White Paper nr. 37 (2012–2013)), next review in 2030.
45. Adopted in 2009 (White Paper nr. 37 (2008–2009)), currently under review, next update in 2025.
46. Adopted 2006, updated in 2011(White Paper nr. 10 (2010–2011), next update in 2020.
47. *See* White Paper nr. 37 (2008–2009).
48. The Convention on Biological Diversity is one of the most central pieces of public international law in the matter.

area. The Management plans also define so-called particularly valuable and vulnerable areas (SVO).

Figure 5.3 Norwegian Management Plan Areas

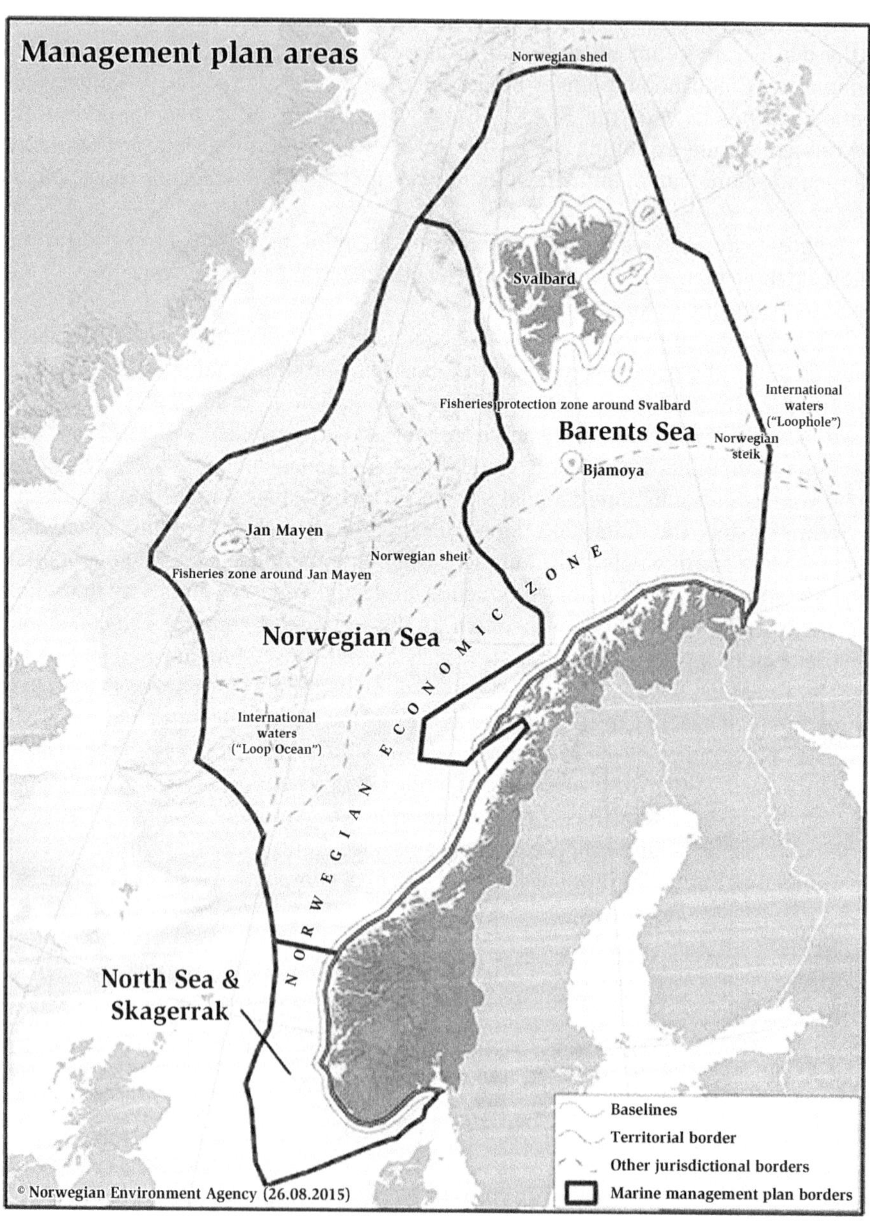

The conflict between uses of the sea areas is a growing concern in Norway. Many commercial and non-commercial interests may enter into conflict. Among them are: petroleum activities, fisheries, shipping, tourism, environmental protection, etc.

Several instruments have been progressively developed, both at a general transversal level and a more sectoral level.

General Principles

The inter-directorate group found that future work on surveying and analyzing the proposed areas must be founded on the Nature Diversity Act's principles of knowledge-based and ecosystem stress; *see* section 8-12 of the Act. On the Continental Shelf of Norway's EEZ, sections 1, 3–5, 7–10, 14–16, 57 and 58 will apply where appropriate.[49]

Separate Tools

In the *Detailed Plan for construction and operation* of the offshore renewable energy installation, it is expected that interaction with other commercial and non-commercial activities are detailed in its impact assessment even more.

Area fee – The establishment of installations in these areas can be made subject to an area fee (OREP Act, section 10-2). The use of the area for the purpose of electricity generation based on renewable energy sources can be subject to an area fee. If so, the area fee is to be paid by the licensee to the State. The MPE can adopt implementation rules for the purpose of that provision.

Pursuant to the Environmental Protection Act, particular attention must be paid to the interaction with (proposed) *conservation areas*, particularly valuable and vulnerable areas (SVOs), seabirds and fish.

[5] International Law and/or European Law

This section raises the question of the applicability of international and European law when deployment of offshore wind power plants takes place in the Norwegian continental shelf or the EEZ.

[a] International Law

The UNCLOS

Under international law, the primary source of legal obligation for developing offshore wind project is the 1982 UNCLOS, which entered into force in 1994 and that Norway ratified in 1996. The OREP Act is applied within the limits defined by the UNCLOS.

49. NVE, Offshore wind power in Norway, Strategic environmental assessment, 2010.

UNCLOS allows Norway to regulate exploitation of renewable marine energy resources. The regulation of this right is performed partly by defining various zones and partly by deciding what rights States have in the various kinds of zones.

UNCLOS divides geographical sea areas into different zones. From the individual state's land territory, inshore waters extend to the baseline (Act relating to Norway's territorial waters and adjacent zone (LOV 2003-06-27 No. 57) sections 1 and 3, confer also Statutory Regulations on the baselines for the maritime territory around Norway proper (FOR 2002-06-14 No. 625), Statutory Regulations on the baselines for the maritime territory around Svalbard (FOR 2001-06-01 No. 556) and Statutory Regulations on the Norwegian maritime territory around Jan Mayen (FOR 2002 -08-30 No. 943)). From the baseline, the territorial waters extend 12 nautical miles out to the EEZ (Act relating to Norway's territorial waters and adjacent zone (LOV 2003-06-27 No. 57) section 2). The EEZ stretches 200 nautical miles from the baseline. Norway has, pursuant to the Act relating to Norway's Economic Zone (LOV 17-12 No. 91) created three zones: (1) an economic zone in the ocean areas off Norway, conferred by Statutory Regulations on implementation of Norway's economic zone (FOR 1976-12-17-15), (2) a fisheries protection zone around Svalbard (FOR 1977-06-03-6), and (3) a fisheries protection zone around Jan Mayen (FOR-1980-05-23-4). The continental shelf also stretches in principle 200 nautical miles from the baseline (Act relating to scientific research and exploration for and exploitation of subsea natural resources other than petroleum resources (LOV 1963-06-21 No. 12)). A state may seek recognition that its continental shelf extends further out; this must be done in accordance with the procedures ensuing from UNCLOS, with submission of documentation from the littoral state and approval from the Continental Shelf Commission. In 2009 Norway achieved recognition of its continental shelf beyond 200 nautical miles in the Barents Sea, the Norwegian Sea and the Arctic Ocean. Beyond the continental shelf is what international law calls the high seas. The Offshore Energy Act uses the above zones in its provisions with regard to geographical applicability.

Since wind is not considered a natural deposit, it is first and foremost the provisions regarding facilities and cables that are of significance in connection with area evaluations under the Offshore Energy Act.

Convention for the Protection of the Marine Environment in the North East Atlantic (OSPAR)

The objective of the OSPAR Convention of September 22, 1992 is to prevent and eliminate pollution and to protect the North-east Atlantic against the harmful effects of human activity. In addition, it shall protect the marine ecosystem and, where practically possible, restore damaged marine areas. The work is led by the OSPAR commission, which consists of representatives of the fifteen contracting countries and the European Commission.

OSPAR has adopted its own guidelines for evaluation of environmental impacts of offshore wind farms (OSPAR Agreement No. 3 2008). Proposition No. 107 to the Odelsting (2008–2009) states that these guidelines will apply to the work of area evaluations, impact statements, license consideration and decommissioning of energy facilities covered by the OREP Act. The guidelines provide guidance about the evaluations that need to be made and the factors that need to be taken into account in five phases of the life-cycle of an offshore wind farm: siting, licenses, monitoring, development and operation and removal/decommissioning. The guidelines describe what biotic and abiotic factors should be emphasized in the choice of areas for offshore wind power, and what other societal interests may be affected and should be considered.

[b] EU/EEA Law

Reminder: Scope of Application EEA Agreement

The geographical scope of the EEA Agreement is defined in its Article 126. It provides that the EEA Agreement applies to the territory of the Kingdom of Norway, but not to Svalbard. The official position of Norway is that the term territory is to be understood in accordance with established practice in international law, which means the EEA Agreement applies to Norwegian land territory, internal waters and territorial waters, but not to the EEZ, the continental shelf or the high seas. This does not prevent Norway to decide, in exceptional circumstances, to accept the incorporation into the EEA Agreement of an EU legislative act which applies to the continental shelf or the EEZ, while reminding that the principle remains.[50]

EU Legislation of Particular Relevance

In terms of targets and permitting rules, the relevant EU legislations incorporated into the EEA Agreement are the Renewable Energy Directive 2009/28/EC and the Electricity Directive.

In terms of support schemes, the EU regime for state aid rules will be applicable, both to the Treaty provisions and European Commission's Guidelines on State aid for environmental protection and energy 2014-2020.

Because of the scope of application of the EEA Agreement, some relevant pieces of EU legislations do not apply in Norway, although sometimes followed in practice. An example is the Marine Strategy Framework Directive (2008/56/EC), which requires Member States to draw up marine strategies (management plans) to achieve good environmental status in their marine areas. The Norwegian government is of the opinion that the Directive covers some geographic areas which fall outside the geographical scope of the EEA Agreement.

50. See Report to the Storting (White Paper), The EEA Agreement and Norway's other agreements with the EU (Meld. St. 5 (2012–2013)), p. 13.

[6] Public Participation Scheme

Besides the SEA and EIA procedures which include consultation of the population and third parties, there is no public participation scheme for offshore wind. Some compensation measures may be provided to the municipalities affected by offshore wind projects, but this is reflected in the license.

[7] Grid

[a] *Rules on Laying of an Undersea Electricity Cable*

Laying down transmission cables from offshore wind power installations to land are subject to a license in accordance with the OREP Act outside the baseline and a license under the Energy Act inside the baseline.

[b] *Existing Pipelines and Cables*

Strategic Impact Assessment 2013 reveals that Utsira nord is the only zone where there are pipelines on the seabed. The pipelines are located in the south-eastern corner of the zone. There are no subsea power cables in any of the zones.

[c] *Connection of Offshore Wind Power to Interconnectors*

An introductory analysis carried out by SINTEF Energy looked at the economics of connecting a wind farm in the south of the Norwegian sector of the North Sea to an interconnector between Norway and Germany, as opposed to a dedicated radial connection from the power station to shore in Norway. Connecting the wind farm to the interconnector would save grid investments. At the same time, such a solution would entail a loss of trade revenues, because wind farms in the south of the Norwegian North Sea will reduce the opportunity of importing from Germany in periods when the prices are lower in Germany than in Norway.

The economic evaluation of whether to build the one or the other alternative rests mainly on whether the reduced import capacity entailed by connecting the wind farms to the interconnector will be balanced by reduced grid investments and lower network losses. Connecting to the interconnector also presupposes an efficient market with rapid market clearing, so that spare capacity not exploited by the wind farms becomes available for trading.

[8] Other Concerns and Legal Regime: From Conflicts in Area Use to Coexistence

A main claimed concern for the future development of offshore wind power in Norway is the interaction with other interests and conflicts in area use. As long as offshore wind is not a commercial activity in Norway, this concern remains hypothetical. But technological innovation and a possible forthcoming more ambitious government policy may turn it to reality. Hereafter are listed the activities offshore wind developments can conflict or interact with in Norway.

[a] Offshore Petroleum Activities

The petroleum industry is the main source of income for the state in Norway, and an important part of the economy in general, with related service and supply industry. Conflicts of interest are identified at the level of the opening of the petroleum areas, licensing procedures and related impact assessments. Possible interaction with offshore wind projects - at the level of e.g., energy supply - would be relevant part of the assessment of the Plan for Development and Operation and of the Decommissioning Plan. A main concern expressed by the petroleum industry is that, once a wind farm has been established, surveying of resources with the aid of seismics and wells will become extremely difficult.[51] The Norwegian Petroleum Directorate nevertheless assumes that the coexistence of wind power and petroleum installations is possible within the most relevant zones (Sørlige Nordsjø I and II, Stadthavet and Frøyabanken).

[b] Shipping

The level of ship traffic activity along the Norwegian coast is very high. Three factors are of especial significance for considering possible areas for wind power: risk traffic, international traffic separation systems and coastal fairways. The Norwegian Coastal Administration has proposed new boundaries for some of the zones due to conflicts with existing leads (the zones Frøyabanken, Træna, Olderveggen and Trænafjorden may have a highest conflict potential). Co-existence between wind power and shipping is possible in most of the areas, but may require alterations of existing leads and rearranging of beacons/lighthouses.

[c] Fisheries

Fisheries represent another important source of economic activities and income for the Norwegian economy. A general feature is that relatively shallow areas that appear suitable for wind power are in many cases also important for fisheries. This suggests that there may be conflicts of interest between the fisheries and wind power in several areas along the coast.[52] Based on today's technologies, the Norwegian Directorate for Fisheries has made a general conclusion that coexistence between wind farms and fishery activities will not be possible within a zone. The Directorate recommends that seven of the fifteen areas are not opened for license applications. These zones are

51. NVE, 2010 Study.
52. NVE, 2010 Study.

Sandskallen – Sørøya nord, Nordmela, Oldervеggen, Frøyagrunnene, Trænafjorden – Selvær, Træna vest and Nordøyan – Ytre Vikna.

Another parallel concern with the offshore petroleum sector is the rules in the OREP Act on the compensation to fishermen in Norway (Chapter 9). The OREP Act contains provisions as to the compensation for economic losses inflicted to fishermen in relation to pollution and waste. The provisions for a compensation scheme for Norwegian fishermen are based on the model of the regulations applicable to the petroleum sector.

[d] Habitat and Species Protection: Biodiversity

In the framework of the SEA work, the inter-directorate group (2010 study) has focused on seabirds, fish and marine mammals. Impacts on seabirds and migrating birds are found to be small to moderate in all zones. Negative impacts of wind farms on fish are found to be small or even non-detectable. Noise from the construction phase can have an impact on Killer whales' behavior, possibly preventing them from hunting close to construction sites (8 km around the construction site).

Impacts on benthic organisms depend on the size and form of the wind turbine foundations. In general, impacts are found to be small to moderate within the fifteen assessed zones.

The relevant legal basis for assessing interaction between offshore wind projects and nature (both habitat and species) is the Biodiversity Act (which replaced the Nature Conservancy Act as of July 1, 2009). The Act covers all nature and all sectors that manage nature or that take decisions that affect nature. The geographical scope of the Act is Norwegian land territory, including lakes and watercourses, plus Norway's territorial waters. On the continental shelf and in Norway's economic zone, the provisions of sections 1, 3–5, 7–10, 14–16, 57 and 58 will apply where appropriate. The object of the Biodiversity Act is to strengthen the protection of nature through sustainable use and protection, so that nature can support human activity, culture, health and well-being, now and in the future, including as a basis for Sami culture. The Act has consistent rules that apply to all sectors. It stipulates management targets for species, biotopes and ecosystems, and enshrines in statute a number of key eco-management principles, including the principle of knowledge-based management, the precautionary principle, the principle of ecosystem approach and overall stress, the principle that the costs of environmental degradation shall be borne by the polluter, etc., and finally the principle of application of environmentally sound techniques and operating methods. These principles shall underpin public decisions that affect biodiversity. In addition to new and updated rules on area protection, including marine protection, the Act contains new policy tools for species protection (prioritized species) and conservation of biotopes (selected biotopes).

[e] Sami Community

The OREP Act provides that the interests of the Sami Community should be taken into due consideration when decisions are taken in accordance with this Act and directly affect their interests. In such case the natural resources forming the basis of the Sami culture shall be taken into consideration. This is consistent with Article 108 of the Norwegian Constitution.

[f] Landscape

Visual impacts from offshore wind turbines on the landscape generally increase the closer the turbines are situated to shore. The visual impact from wind turbines is not considered to be unacceptable in any of the fifteen zones selected in the SEA. However, it is often a main ground for local municipalities to refuse offshore wind project within their area of competence (within baseline), due to strong local opposition of the population as well as actors of the tourism and outdoor recreation sectors.

[g] National Security-Related Regulations, Defense Concerns and Military Activities

The Armed Forces have interests related to training areas, radar facilities and other infrastructure. The zones Gimsøy nord and Utsira nord overlap with areas used by the Norwegian Air Force and the Norwegian Navy for practice purposes, and development of wind power will be in direct conflict with today's use of the areas. In most cases the opportunity for coexistence will have to be clarified ad hoc when the siting and size of any wind farms is known.[53]

§5.04 CHALLENGES AND SOLUTIONS

[A] Institutional Design: Challenges and Solutions

[1] Challenges

There are few apparent institutional challenges to the development of offshore wind in Norway. Similarly, there is a large political consensus at national level for the development of offshore wind, as long as it is economically viable and do not contradicts with other industries and interests. Meanwhile, the lack of concrete targets is slowing down the development of offshore wind power, while local municipalities may not support the projects due to local opposition. In conclusion, the real challenges are more technical and economical.

53. NVE, 2010 Study.

[2] *Solutions*

The Parliament requested the government to put forward a strategy – including support measures – for the development of offshore wind power within 2017. Additionally, in the state budget for 2017, the government stated that it would deliver a strategy for the commercial development of floating wind power that could contribute to the electrification of the Norwegian continental shelf. This is expected to be done in the state budget for 2018. Defining a clear policy at state level will contribute actively to the development of the sector. At local level, the adoption of public participation measures and schemes, including compensation or interest schemes, could be a manner to solve the problem of local opposition, although Norwegian coast population and the tourist industry is strongly attached to costal landscape protection.

[B] Incentives: Challenges and Solutions

[1] *Challenges*

The already-high ratio of renewable energy production combined with concerns about wind power's local environmental impacts has provided fuel for considerable public debate in Norway in recent years.

In parallel, there is a knowledge gap which works as hinder to sound decisions. The inter-directorate group found that there will be a need for preliminary and follow-up studies of important topics, such as birds and fish, if offshore wind power is to be developed in Norway. This will be required to clarify real impacts and increase the knowledge base for subsequent developments.[54]

[2] *Solutions*

The adoption of targets and specific incentives for offshore wind power, both bottom-fixed and floating – could be instrumental in supporting the sector in addition to R&D programs.

There is an obvious need for the collection of more data as to the effects of offshore wind developed on the local environment.

[C] Regulations: Challenges and Solutions

[1] *Challenges*

In certain cases, the interactions between the Energy Act and the OREP Act may need to be clarified, such as for connection to shore.

54. NVE, 2010 Study.

A challenge is that the applicable regime may need to be clarified for each individual project separately. The establishment of wind farms at sea must be clarified with reference to other relevant legislation such as the Energy Act, the Harbours and Fairways Act, the Planning Act, the Ocean Resources Act, the Petroleum Act and the Cultural Heritage Act. The various acts have different application areas and the relationship to relevant legislation must therefore be clarified specifically for concrete wind power projects.

The Norwegian regime for offshore wind outside the baseline has many similarities with the Norwegian petroleum regime in its approach and basic principles and tools. Transparency and predictability are among the core qualities of that regime. However, many implementation aspects must still be adopted in the case of offshore wind, not least due to a lack of practice and clear strategy from the government.

The OREP Act provides a framework for the regulation of renewable energy production at sea. On a number of points, the Act must be supplemented by statutory regulations and provisions.

[2] Solutions

Again, one possible solution could be the adoption of clear targets for the building of offshore wind power capacity generation, at domestic level but also in connection with the cluster approach between offshore infrastructures as promoted in the North Sea.

The regulatory framework could with advantage enable for greater cooperation between the different commercial activities at sea. A prerequisite is the elaboration of a good knowledge basis on the effects of offshore wind in the relevant areas, and the completion of sound SEAs and EIAs.

In absence of practice, many provisions are subject to interpretation, including the licensing terms for offshore wind installations. The OREP Act gives the possibility for the MPE to adopt implementation legislation in the form of Regulations including on the opening of licensing areas, the licensing terms, the terms of the Detailed Plan for construction and operation of the installation. If the government aims to ensure visibility and transparency, equal treatment of the demands, the government should pursue to the adoption of such legislation.

A system of prequalification could be established to speed up the licensing procedure, as it is already applied under the Petroleum Act. Section 3-3 of the OREP Act opens for such possibility.

There is an important need for harmonization of implementation measures. The OREP Act provides the Ministry the competence to do so.

Finally, the national strategy for offshore wind may be to an increasing extent be influenced by regional cooperation and the development of North Sea wind power grid network or "cluster" approach.

§5.05 CONCLUSION

Discussing offshore wind energy in Norway is both a question of energy and industrial policy. From an energy and environmental perspective, it raises the question of the need to further diversify the energy mix. Offshore wind can with difficulties compete with hydropower at the moment. However, there are strong signals in favor of offshore wind development in the mid- to long term.

The positive part is that the legislation is in place. The negative part is that it is not used (OREP Act) or with difficulties (Energy Act). The current main support scheme for new renewable energy sources is not adapted to a technology which is still non-competitive in that region. R&D support has enabled the development of promising technologies, which need to be demonstrated and sold.

Norway has consequently taken a series of regulatory steps in the right direction, and can benefit from the expertise of its offshore industries, but still lacks the momentum to let its nascent offshore wind industry blossom.

PART III Offshore Wind Power of Developing Countries in the Pacific Asia

CHAPTER 6

Legal Framework to Develop Offshore Wind Power in United States

Jeremy Firestone

§6.01 INTRODUCTION

European nations have been installing offshore wind turbines since 1991, providing approximately 11,000 MW of clean energy.[1] As early as 2005, it appeared the United States (US) would follow Europe's lead when the US Congress enacted section 388 of the Energy Policy Act of 2005, authorizing the US Department of the Interior (US DOI) to lease federal waters (generally oceanic waters 3-200 nautical miles from the US coast) for offshore wind power development. While the US had almost 74,000 MW of wind power installed on land at the end of 2015,[2] it had yet to install its first commercial scale offshore wind turbine. US Offshore Wind Power has for the most part been a missed opportunity.[3] This is particularly so given the large offshore wind resource; indeed, the resource of the east coast of the US alone is large enough to meet the US' electric, auto passenger transport and building heat needs.[4]

---

1. EWEA, The European Offshore Wind Industry – Key Trends and Statistics 2015 (2016), available at https://windeurope.org/wp-content/uploads/files/about-wind/statistics/EWEA-European-Offshore-Statistics-2015.pdf (last visited July 25, 2016).
2. AWEA, US Wind Industry First Quarter 2016 Market Report (2016), available at http://awea.files.cms-plus.com/FileDownloads/pdfs/1Q2016%20AWEA%20Market%20Report%20Public%20Version.pdf (last visited July 25, 2016).
3. Jeremy Firestone, Cristina A. Archer, Meryl P. Gardner, John A. Madsen, Ajay K. Prasad, Dana E. Veron, The time has come for offshore wind power in the US, *Proceedings of the National Academy of Sciences*, 112(39): 11985–11988, doi: 10.1073/pnas.1515376112 (2015).
4. Kempton W, Archer CL, Garvine RW, Dhanju A, Jacobson MZ (2007) Large CO2 reductions via offshore wind power matched to inherent storage in energy end-uses *Geophys Res Lett* 34: L02817. doi:10.1029/2006GL028016.

There are nevertheless reasons for optimism: The US has leased large areas of the ocean to offshore wind power developers,[5] including entities such as DONG Energy, the most experienced offshore wind developer in Europe, and hence the world; the State of Massachusetts has adopted legislation requiring energy utilities to secure 1,600 MW of offshore wind power;[6] prices have been falling dramatically in Europe for new offshore wind projects, which bodes well for a US market affected by low natural gas prices;[7] a 30 MW demonstration-scale project was commissioned off on Block Island, Rhode Island in late 2016; the State of Maryland approved two wind contracts for a total of 368 MW and New York's Long Island Power Authority approved a contract for 90MW;[8] US tax policy is positioned in a manner such that it may well shift toward offshore wind power support; the US has adopted its Clean Power Plan, which will result in greenhouse gas (GHG) emission reductions from the electricity sector of 32% by 2030;[9] the leaders of the US, Mexico and Canada recently pledged to achieve 50% clean power generation by 2025;[10] and offshore wind power of the US' Pacific coast is primed to take off in California[11] given aggressive state mandates for renewables and Hawaii,[12] which generates most of its electricity from imported petroleum and as a consequence, has very high electricity rates.[13]

Although much uncertainty exists at the federal level given the recent US presidential election of Donald Trump, particularly as it relates to the Clean Power Plan,[14] with the existing offshore leases along the Atlantic coast, most of the near-term

---

5. EIA, Federal Leasing for offshore wind grows as first U.S. offshore wind farm comes online (December 30, 2016), http://www.eia.gov/todayinenergy/detail.php?id=29372 (last visited January 3, 2017).
6. An Act Relative to Energy Diversity, H. 4568, available at https://malegislature.gov/Bills/189/H4568.pdf (last visited January 3, 2017).
7. *See* e.g., https://www.government.nl/latest/news/2016/12/12/dutch-consortium-to-construct-second-borssele-offshore-wind-farm (last visited January 3, 2017).
8. http://dwwind.com/project/block-island-wind-farm/ (last visited January 3, 2017); Order No. 88192, In The Matter of the Applications of US Wind, Inc. and Skipjack Offshore Energy, LLC for a Proposed Offshore Wind Project(s) Pursuant To the Maryland Offshore Wind Energy Act of 2013, Docket No. 9431. Maryland Public Service Commission (May 11, 2017), http://webapp.psc.state.md.us/newIntranet/casenum/NewIndex3_VOpenFile.cfm?ServerFilePath=C:\Casenum\9400-9499\9431\\121.pdf (last visited August 18, 2017); https://www.governor.ny.gov/news/governor-cuomo-announces-approval-largest-offshore-wind-project-nation (last visited August 18, 2017).
9. US EPA, Carbon Pollution Emission Guidelines for Existing Stationary Sources: Electric Utility Generating Units, 80 Fed Reg 64662 (October 23, 2015). The US Supreme Court has since stayed implementation of the Clean Power Plan pending judicial review on its merits. *West Virginia v. EPA*, 15A773, __ US ___ (S. Ct. February 9, 2016), available at https://www.supremecourt.gov/orders/courtorders/020916zr4_4g15.pdf (last visited January 3, 2017).
10. The White House, Leaders' Statement on a North American Climate, Clean Energy, and Environment Partnership (June 29, 2016), available at https://www.whitehouse.gov/the-press-office/2016/06/29/leaders-statement-north-american-climate-clean-energy-and-environment (last visited July 25, 2016).
11. http://www.boem.gov/California/ (last visited July 28, 2016).
12. http://www.boem.gov/Hawaii/ (last visited July 28, 2016).
13. Institute for Energy Research, Hawaii Energy Facts, http://instituteforenergyresearch.org/media/state-regs/pdf/Hawaii.pdf (last visited July 25, 2016).
14. https://www.epa.gov/newsreleases/epa-review-clean-power-plan-under-president-trumps-executive-order (last visited August 18, 2017).

policy support will be required of the States (like Massachusetts recent legislation noted above) rather than the federal government. Indeed, Statoil's entry into the US offshore wind market post-election with a record shattering bid to lease the New York Wind Energy Area[15] evinces confidence in state policy and perhaps Donald Trump's commitment to investing in infrastructure as well.

This chapter traces the institutional framework for offshore wind power in the US, including the primary regulatory framework, the wider ocean management framework in which it resides, the requirements for environmental assessment, and the role of the public in the process.

§6.02 INSTITUTIONAL DESIGN FOR OFFSHORE WIND

[A] The "Main" Authority During the Development of Offshore Wind Power

Section 388 of the Energy Policy Act placed offshore wind power development primarily in the hands of US DOI. Within US DOI, the offshore wind leasing and construction program is administered by its Bureau of Ocean Energy Management (BOEM);[16] however, as will be detailed below, there is a crazy quilt of federal agencies engaged in various aspects of ocean management.[17] Section 388 amended the Outer Continental Shelf Lands Act (OCSLA),[18] which prior to that time had primarily regulated offshore oil and gas.

While BOEM leases federal waters for offshore wind power, it primarily functions as a regulatory agency as compared to an advocate for offshore wind power, and thus, is generally agnostic as to the development of offshore wind energy. In contrast, the US Department of Energy (US DOE), with whom BOEM coordinates closely, has assumed the promotion mantle. In its 2008 report, US DOE envisioned 54 GW of offshore wind power by 2030;[19] a goal that has now been scaled back to the still aggressive, and yet unrealistic, 3 GW by 2020, 22 GW by 2030 (with a more robust 86 GW by 2050) in its 2015 Wind Vision.[20] More recently, US DOE and BOEM issued a joint offshore wind

15. Interior Department Auctions Over 79,000 Acres Offshore New York for Wind Energy Development, https://www.doi.gov/pressreleases/interior-department-auctions-over-79000-acres-offshore-new-york-wind-energy (last visited January 3, 2017).
16. BOEM along with the Bureau of Safety and Environmental Enforcement (BSEE), were previously in one US DOI division known as the Minerals Management Service (MMS), but were renamed and separated after the BP Deepwater Horizon Oil Disaster in the Gulf in 2010.
17. *See* Jeremy Firestone, Willett Kempton, Andrew Krueger and Christy L. Loper, Regulating Offshore Wind Power and Aquaculture: Messages from Land and Sea, *Cornell Journal of Law and Public Policy*, 14(1): 71–111 (2004) *republished in Environmental Law Reporter* (Environmental Law Institute), 35:10289–10307 (May 2005).
18. 43 U.S.C. §§ 1331 et seq.
19. DOE, 20% Wind Energy by 2030 Increasing Wind Energy's Contribution to U.S. Electricity Supply, DOE/GO-102008-2567 July 2008, available at www.nrel.gov/docs/fy08osti/41869.pdf (last visited July 25, 2016).
20. DOE, http://www.energy.gov/sites/prod/files/WindVision_Report_final.pdf (last visited July 25, 2016).

power "strategy,"[21] which focuses on cost reduction, stewardship, and increasing understanding of offshore wind's benefits and costs.

[B] The "Supplementary" Authorities During the Development of Offshore Wind Power

[1] Central Government Level

Unlike some countries such as Canada that have more integrated ocean management, the US has twenty-two federal agencies that focus on narrow single issue mandates rather than an integrated whole, although that is changing slowly. Authorities can be regulatory or research in orientation. The primary regulatory agencies and their authorities are as follows:

- US Coast Guard (USCG) (commercial shipping fairways and traffic separation schemes[22]).
- US Army Corps of Engineers (USACE) (obstructions to navigation; dredging and filing; sand mining).[23]
- National Oceanic and Oceanographic Administration (NOAA) (marine mammals,[24] endangered species,[25] sea turtles,[26] commercial and recreational fisheries,[27] marine aquaculture[28]).[29]
- US Fish and Wildlife Service (USFWS) (migratory birds,[30] marine mammals, endangered species, sea turtles).
- US Environmental Protection Agency (USEPA) (air discharge from service vehicles;[31] climate[32]).
- US Navy (training and testing; submarine cables[33]).

21. DOE and BOEM, National Offshore Wind Strategy: Facilitating the Development of the Offshore Wind Industry in the United States, https://energy.gov/sites/prod/files/2016/09/f33/National-Offshore-Wind-Strategy-report-09082016.pdf (last visited January 3, 2017). The "strategy" is more a framework than a plan of how to actually achieve the aforementioned targets.
22. Ports and Waterways Safety Act (PWSA), 33 U.S.C. 1223.
23. *See* in particular, Rivers and Harbors Act of 1899 (RHA), 43 USC § 403.
24. Marine Mammal Protection Act (MMPA), 16 U.S.C. §§1361 et seq.
25. Endangered Species Act (ESA), 16 U.S.C. §§1531 et seq.
26. In the US, species and threatened and endangered and are protected under the ESA.
27. Magnuson-Stevens Fishery Conservation and Management Act, 16 U.S.C. §§1801 et seq.
28. Fisheries of the Caribbean, Gulf, and South Atlantic; Aquaculture, Final Rule, 81 Fed Reg 1762-1800 (January 13, 2016) to be codified at 50 CFR Part 662.
29. NOAA also has nonregulatory responsibilities for weather and climate.
30. Migratory Bird Treaty Act (MBTA), 16 U.S.C. §§703 et seq.
31. Clean Air Act (CAA), 42 U.S.C. § 7627; 40 CFR Part 55.
32. CAA, 42 U.S.C. § 7411.
33. Naval Sea Floor Cable Protection Office, http://www.navfac.navy.mil/products_and_services/ci/products_and_services/naval_ocean_facilities_program/sea_floor_cable_protection_nscpo.html (last visited July 26, 2016).

In addition there are other federal agencies that bear mention because their jurisdiction can affect and can be affected by the development of offshore wind power. These include:

- Federal Energy Regulatory Commission (FERC) (transmission and the sale of electricity at the wholesale level;[34] marine hydrokinetics[35]).
- Federal Aviation Administration (FAA) (aircraft obstructions[36]).
- National Park Service (NPS) - management of national seashores and lakeshores, with a mandate to "conserve the scenery"[37].

Finally, US DOE, the BOEM Environmental Studies Program, and NOAA's Sea Grant program play an important role by funding research to study various aspects of offshore wind power (technical, environmental, social, financial, demonstration, grid integration).

[2] The Role of the State Government

In the US, state governments play a critical role in the development of offshore wind power. To begin with, states control the development of natural resources within the first three nautical miles from shore.[38] That has two implications: First, any cable transmitting energy from an offshore wind project located in federal waters must pass through the submerged lands controlled by a state. Second, although the federal government is not without authority in state waters, a state would be the lead agency to decide whether or not to develop its state waters for offshore wind power. Good examples are Deepwater Wind's 30 MW Block Island, Rhode Island demonstration project,[39] and Fishermen Energy's Atlantic City project.[40] In a similar vein, states control the development of inland waters, including bays (such as Delaware Bay and Chesapeake Bay) and the Great Lakes, with the Icebreaker Offshore Wind Power Project outside of Cleveland, Ohio[41] being the furthest along. Moreover, under the Public Trust Doctrine, states control the lands subject to the ebb and flow of the tide and thus have to ensure that any cabling will not impair the public trust.[42]

In addition, states have authority under federal government's Coastal Zone Management Act (CZMA).[43] A primary federal goal of the CZMA was to have states adopt coastal management plans. The carrot, in pertinent part, was that a state that adopted such a plan would then have authority to determine whether or not a federal

34. Federal Power Act, 16 USC § 824.
35. In re Verdant Power LLC, 111 FERC ¶ 61,024 (2005), order on reh'g, 112 FERC ¶ 61,143 (2005).
36. 14 CFR Part 77.
37. National Park Service Organic Act of 1916, 54 U.S.C. § 100101.
38. Submerged Lands Act of 1953, 43 U.S.C. §§ 1301 et seq. Florida on its Gulf Coast and Texas control large portions as does Puerto Rico.
39. http://dwwind.com/project/block-island-wind-farm/ (last visited July 25, 2016).
40. http://www.fishermensenergy.com/atlantic-city-windfarm.php (last visited July 25, 2016).
41. http://www.leedco.org/ (last visited July 25, 2016).
42. *Phillips Petroleum Co v. Mississippi*, 484 US 469 (1988).
43. 16 U.S.C. § 1451 et seq.

authorization (e.g., lease or permit) that might affect land (both submerged and fastlands), water, or natural resources subject to the plan was consistent with that plan.[44] Absent a rare federal override,[45] a proposed authorization deemed inconsistent by a state with its plan could not move forward. Thus, if a state were to prohibit offshore wind power in state waters as part of its state plan, that state would have a strong case for finding such a project in federal waters to be inconsistent.

States also play a large role when it comes to energy facility siting, and approval of contracts for the purchase of electricity, renewable energy mandates, and GHG emission reduction requirements. For example, the State of Massachusetts Energy Facility Siting Board has jurisdiction, M.G.L. c. 164 §69J1/4 over any proposed generation facility greater than 100 MW. Power purchase agreements (PPA) between a developer and an electric service provider (utility) are also subject to approval by the state in question, though a public utility commission. For example, after consideration, the Delaware Public Service Commission approved the first PPA for offshore wind power in the Americas in 2008,[46] although the developer later abandoned the PPA due to financing difficulties.

State Renewable Portfolio Standards (RPS) also play an important role because they mandate that a certain percentage of an electric utility's delivered energy be from renewable sources or alternatively that the utility pay a compliance penalty. Evidence of compliance is gained by holding renewable energy certificates (RECs), each of which are equal to one megawatt-hour (MWh) of renewable energy production. Hawaii is the most aggressive state in that regard, adopting in 2015 HB623,[47] which puts it on a path to be 100% reliant on renewable energy sources by 2045. Other states like Maryland have an offshore wind specific requirement or OREC, which effectively creates an offshore wind power set aside.[48] Finally, a number of states have GHG reduction standards. In that regard, California is the most notable, with its plan of achieving 40% below 1990 levels by 2030 and 80% by 2050.[49]

[3] Other Players Vital for the Development of Offshore Wind

There are a number of other entities that have played and continue to play a vital role in the development of offshore wind power in the US. On the regulatory side are Regional Transmission Organizations (RTOs) and Independent System Operators (ISOs), which provide nondiscriminatory access and interconnection to the grid; wholesale sale of electricity; regulation of demand response at the wholesale level; and

44. 16 U.S.C. § 1456(c)(3).
45. *Id.*
46. In the Matter of Delmarva Power Company, PSC Docket No. 06-241, Order No. 7440 (2008) available at http://depsc.delaware.gov/orders/7440.pdf (last visited July 25, 2016).
47. http://www.capitol.hawaii.gov/session2015/bills/HB623_CD1_.pdf (last visited July 25, 2016).
48. Maryland Code, §§ 7-701 et seq.
49. Executive Order B-30-15 (April 29, 2015), https://www.gov.ca.gov/news.php?id=18938 (last visited July 25, 2016).

frequency regulation. Examples include PJM, which covers the mid-Atlantic, ISO New England, and NYISO.[50]

In the US, Indian tribes also are important governmental entities when it comes to offshore wind power. Indian Tribes are one of the four types of governmental entities mentioned in the US Constitution (the federal government, state governments, and foreign governments being the other three), and it has long been held, that tribes, like states (albeit different), have aspects of sovereignty,[51] and as a result of that sovereignty as well as federal preemption of state laws, Tribes have regulatory jurisdiction within the boundaries of their lands, including over non-Indians.[52] Tribes, particularly in the Northwest (such as those with reservations located within the boundaries of the State of Washington) have treaty rights to fish, and tribal cultural and historic resources are protected under US law. Even in the east, where Indian Tribes are much less prevalent, concerns have arisen at the Cape Wind site in Cape Cod, Massachusetts over the effect that project would have on Indian religious practices[53] and regarding the effect on cultural resources of cable-laying to support the Block Island site.[54]

The National Renewable Energy Lab (NREL) plays an important role along with several other national energy labs (such as Lawrence Berkeley and Sandia) in advancing offshore wind power through research and testing. NREL operates the National Wind Technology Center in Boulder, Colorado, which includes a turbine test site; as well, NREL plays an important role in the management of the Wind Technology Testing Center, which is located in Massachusetts, and which can test turbine blades up to 90 m in length). There also is a drivetrain test facility in South Carolina.[55]

Because offshore machines are so large – for example, the blades on the Vestas V8 are 164 m long, thus making it difficult to manufacture key components in locations that are not adjacent to the oceans or other large waterways connected thereto, and because there may be significant economies that can be realized from quayside assembly of turbines (tower, rotor, nacelle, and blades), port development will also be important part of the US' offshore wind power infrastructure strategy. Finally, independent research arms, like the University of Delaware's Center for Carbon-free Power Integration, can help lower social, regulatory, economic, and technical barriers.[56]

---

50. *See* http://www.ferc.gov/industries/electric/indus-act/rto.asp (last visited July 28, 2016).
51. *Worcester v. Georgia*, 31 US 515 (1832).
52. For limits on that jurisdiction *see Montana v. United States*, 450 US 544 (1981).
53. *Public Employees for Environmental Responsibility v. Beaudreau*, 25 F. Supp 3d 67 (D. DC 2014), rev'd on other grounds, *Public Employees for Environmental Responsibility v. Hopper*, ___ F.3d___ (DC Cir. 2016). *See* Jeremy Firestone and Jeffrey Kehne, Wind Energy, Ch. 16, at 381-382 in Michael Gerrard (ed.), The Law of Clean Energy: Efficiency and Renewables, American Bar Association (2011).
54. Providence Journal, Judge Denies Narragansett tribe's bid to suspend construction of cable for wind farm, http://www.providencejournal.com/article/20160516/NEWS/160519484 (last visited July 25, 2016).
55. http://www.nrel.gov/nwtc/ (last visited July 25, 2016); http://www.masscec.com/wind-technology-testing-center (last visited July 25, 2016); http://clemsonenergy.com/ (last visited July 25, 2016).
56. *See* www.carbonfree.udel.edu (last visited July 25, 2016); for Wind Power Program publications, *see* http://www.ceoe.udel.edu/research/affiliated-programs/wind-power-program/publications (last visited July 25, 2016), for the University's research turbine *see* http://www.ceoe.udel.edu/lewesturbine/ (last visited July 27, 2016).

[C] The Role of "Law" under the Institution Decision Making and Related Policies

Law changes very slowly in the US. Part of this is deliberate with its system of checks and balances, but more recently the country has witnessed gridlock, resulting in deadlocked on many fronts, including climate, gun control, abortion, immigration, etc. While wind power in the US could have easily been swept into this dichotomy, wind power has for the most part enjoyed bipartisan support, perhaps owing to the fact that the so-called red states – those with more republican leanings – tend to be the windier land states. Yet, being able to get a bipartisan affirmative vote of support for tax credits for wind power has never been easy, and will likely be more difficult with offshore wind power given its more "blue" state orientation.

There also has been a clear reluctance on the part of both the Bush and Obama Administrations to seek legislative fixes to the offshore wind licensing regime despite the regime's obvious faults, as will be explained in more detail later in this chapter. Perhaps again, this is owing to the extreme difficulty of getting anything accomplished legislatively, particularly, since the midterm November 2010 election in the first Obama term. It also may be owing to a related decision to place political capital elsewhere given the high cost of offshore wind energy relative to low natural gas prices in the US.

§6.03 LEGAL DESIGN FOR OFFSHORE WIND

[A] Incentives

At the federal level, the main financial driver of wind power development on land has been the production tax credit (PTC) and more recently, its alternative, the Investment tax credit (ITC). The PTC, as its name suggests, is a per-kWh PTC while the ITC is a credit based on a percentage of capital cost expenditures. Projects that are more capital intensive, like offshore wind, stand to benefit more from the latter than the former. The problem with both of these tax credits is that they have tended to be in place for a year, or at most two, and have then required Congressional re-authorization. While awaiting re-authorization, tax credits have often lapsed only to be put in place at the end of a tax year or even later and made retroactive. This has resulted in very uneven development year to year in the US. And with a product like offshore wind power, which has a development time horizon of five to seven years, such unevenness can be fatal. In contrast, in section 1306 of the Energy Policy Act of 2005, Congress adopted a nuclear PTC, which provided that a new nuclear plant need not be in service until the end of 2020 to qualify for the tax credit. Senators Carper (D-DE) and Collins (R-ME) came up with a different way around the long lead time conundrum while providing the government with a cap on the impact to the federal treasury by limiting their offshore wind ITC bill to only the first 3000 MW.[57]

More generally, in late 2015, greater policy-sanity reigned, and Congress and the industry came to terms on a five-year wind power PTC with a phasedown in recognition of the increasing cost parity of land-based wind. This opens an opportunity for more narrowly targeted efforts such as the Carper-Collins bill.

US federal loan guarantees also can play an important role in reducing the cost to produce offshore wind energy given the capital-intensive nature of the offshore wind power development by lowering the cost of capital. This tool has unfortunately not been deployed as much as it might. For example, Cape Wind received a loan guarantee for only a minor portion of its project[58] while the demise of the NRG Bluewater Wind project can be traced in part to the federal government's failure to provide greater certainty in that regard.[59]

Given that offshore wind power turbines generate near-carbon-free electricity - there are however emissions related to O&M activities from service vessels and in the manufacture and installation of the turbines - offshore wind power stands to benefit from those jurisdictions that have carbon-trading regimes. This is true, not because of any linkages between the two, but because unlike fossil fuel plants, offshore wind power operators need not pay for carbon allowances. Examples of such relevant carbon schemes are the Regional Greenhouse Gas Initiative (RGGI), which includes Massachusetts, Rhode Island, New York, Delaware and Maryland,[60] and the California Cap and Trade Program[61] (which is has been linked with Quebec for some time, and was linked with Ontario as of January 1, 2017[62]).

Despite state support, there is reason for concern that some state incentives could run afoul of trade laws either internationally or domestically. Here, I concentrate on the domestic front, where the negative implications of the US Constitution's grant of authority to Congress to regulate commerce among the states (the so-called Dormant Commerce Clause[63]) - that is, that the states lack such power, suggests that state authority to favor in-state renewable generation has limits. As a result, for the most part, states have allowed in-state electric providers to meet RPS requirements through

57. Incentivizing Offshore Wind Power Act, http://www.carper.senate.gov/public/index.cfm/pressreleases?ID=4be57e26-3f1e-41e7-807f-ed1f60ab3809 (last visited July 25, 2016). The nuclear PTC was also limited to the first 6000MW.
58. http://energy.gov/articles/energy-department-offers-conditional-commitment-cape-wind-offshore-wind-generation-projec-0 (last visited July 28, 2016).
59. http://nawindpower.com/nrg-energys-bluewater-wind-puts-the-brakes-on-offshore-wind-farm (last visited July 28, 2016). The federal government has also invested in small-scale offshore wind power demonstration projects, through a grant program administered by US DOE. These grants, while substantial (upwards of USD 47M), are tied to specific projects and thus will not be of assistance going forward.
60. www.rggi.org (last visited July 25, 2016).
61. http://www.arb.ca.gov/cc/capandtrade/capandtrade.htm (last visited January 3, 2017).
62. https://www.ontario.ca/page/cap-and-trade (last visited January 3, 2017).
63. *See* US Constitution, Art. 1, Section 8, Clause 3, granting power to Congress to "regulate Commerce with foreign Nations, and among the several States, and with the Indian Tribes."

the purchase of RECs on a regional rather than a state-basis.[64] The inability to favor in-state generation causes a dilemma for states in the offshore wind power context given that at present offshore wind power is priced above market rates. The OREC states of New Jersey and Maryland have sought to work around the Dormant Commerce Clause by requiring any offshore wind power project to demonstrate net economic benefits to the state, which incentivizes developers to use in-state labor, equipment and staging grounds.[65] States also provide tax incentives to business such as to turbine supply companies to locate nacelle assembly plants and blade manufacturing plants in their states.

In 2016, the 8th Circuit Court of Appeals in *North Dakota v. Heydinger*,[66] struck down a Minnesota law that would have prohibited among other things, Minnesota utilities from entering into PPAs for electricity generated out-of-state at newly-constructed coal-fired power plants. However, California has been allowed to establish a related low-carbon fuel standard that affects out-of-state producers because it did so evenhandedly and did not discriminate against out-of-state (or out-of-country – Brazil) producers of biofuels when it considered alternative sources of electricity used to produce a given biofuel and the transportation of that fuel to California.[67] Together, these cases suggest that with careful drafting, states may be able achieve in-state objectives.

[B] Regulations

BOEM implemented section 388 of the Energy Policy Act of 2005[68] through the promulgation of the Renewable Energy Program Regulations, 30 CFR 585, in 2009. The regulations govern planning, leasing, site assessment and construction and operations, with a lease providing site control and sufficient legal control to a developer to satisfy lending institutions and equity players. The regime covers a lease and a permit to build and operate an offshore wind power project and to cable through federal waters. A developer needs separate permissions to cable through state waters and to interconnect to the land-based grid, and in some jurisdictions, permission under a state energy facility siting law. If competitive interest in a lease area is demonstrated, BOEM will hold a multi-stage auction with the lease going to the qualified entity that bids the most. As of the end of 2017, BOEM has leased areas off of Virginia, Maryland, Delaware, New Jersey, New York, Rhode Island, and Massachusetts and is at various

64. In 2010, the State of Massachusetts was sued by TransCanada regarding provisions that would have required electric suppliers to either purchase solar RECs (SRECs) locally or pay a compliance fee and that long-term contracts be negotiated from in-state providers. The case was resolved without a ruling on the merits as Massachusetts grandfathered in some recently signed contracts. *See* http://archive.boston.com/business/technology/articles/2010/05/29/deal_reached_in_state_energy_suit/ (last visited July 25, 2016).
65. In re Fishermen's Atlantic City Wind Farm, A-3932-13T3 (Superior Court of NJ, Appellate Division, May 29, 2015) (rejecting Fishermen's Energy appeal regarding net benefit calculation).
66. 825 F.3d 912 (8th Cir. 2016). Notably, however, the three judges on the panel differed in their reasoning, with only Judge Loken relying on the dormant commerce clause.
67. *Rocky Mountain Farmers Union v. Corey*, 730 F.3d 1070 (9th Cir. 2013).
68. 43 U.S.C. § 1337(p).

stages of the lease development process in California, Hawaii, North Carolina, and South Carolina.[69]

National Environmental Policy Act of 1969 (NEPA), 42 USC 4332, is the law that governs what is more generally referred to throughout the world as "environmental impact assessment." Nomenclature is a bit different in the US, with the US employing the term "environmental impact statement" or "EIS" for a full NEPA evaluation, but the general process should be familiar.[70] NEPA is not substantive - that is, it does not mandate a given result to be compliant; rather it imposes an evaluation process. As long as the environmental consequences of a proposed action are adequately considered, a proposed project is NEPA-compliant. In short, for any major federal action authorized, funded or undertaken by a federal agency,[71] that agency is required to analyze the environmental impacts of the proposed action, including adverse effects that cannot be avoided, and alternatives to the proposed action. Effects may be ecologic, aesthetic, historic, cultural, economic, social or health related.[72] NEPA is beneficial in that it also drives cross-government agency consultation so that agencies such as BOEM receive needed input from other federal agencies, states and Indian tribes on the effects construction of a proposed wind project may have on wildlife, fisheries, marine transportation and the like. NEPA encourages Strategic Environmental Assessment through the undertaking of Programmatic EIS or EIA tiering. BOEM undertook such a programmatic EIS in 2007 related to offshore alternative energy development (wind, wave, tidal and ocean current).[73]

The White House Council on Environmental Quality (CEQ) has promulgated NEPA regulations[74] that are applicable across the federal government; agencies such as US DOI have, in addition, their own-agency specific regulations detailing how they will implement NEPA, to which they also must adhere.[75] Although typically, US laws do not have extraterritorial effect, NEPA has been held to apply in the US Exclusive Economic Zone (EEZ).[76]

---

69. *See* http://www.boem.gov/Renewable-Energy-State-Activities/ (last visited January 3, 2017).
70. Sometimes, a briefer document, an Environmental Assessment (EA) is prepared. An EA seeks to answer the question of whether the effects may be "significant." If they are not, the agency will issue a FONSI (Finding of No Significant Impact). If they are, they agency is obliged to prepare an EIS. A very small proposed action can receive categorical exclusion (CE) from NEPA but even the University of Delaware's single 2MW land-based wind turbine did not receive a CE so one should not expect a CE in the offshore context. *See* http://energy.gov/nepa/ea-1782-university-delaware-lewes-campus-onsite-wind-energy-project (last visited July 25, 2016).
71. 43 CFR 1508.18.
72. 43 CFR 1508.08.
73. Programmatic Environmental Impact Statement for Alternative Energy Development and Production and Alternate Use of Facilities on the Outer Continental Shelf (2007), available at http://www.boem.gov/Renewable-Energy-Program/Regulatory-Information/Guide-To-EIS.aspx (last visited July 25, 2016).
74. 40 CFR Parts 1500–1508.
75. *See* https://www.doi.gov/oepc/HQ-teams/nrm-team (last visited July 25, 2016).
76. *NRDC v. US Dept. of Navy,* 2002 *Lexis* 26360 (C.D. CA 2002), available at https://coast.noaa.gov/data/Documents/OceanLawSearch/Natural%20Resources%20Defense%20Council%20v.%20U.S.%20Department%20of%20Navy,%20No.%20CV-01-07781%20CAS%20(RZX),%20(C.D.%20Cal.%202002).pdf (last visited July 26, 2016).

In litigation over an EIS, litigants typically are focused on the question of whether or not the EIS is adequate. Much litigation has arisen out of the Cape Wind project, proposed to be located in Nantucket Sound off the coast of Cape Cod, Massachusetts, including on the EIS; as well as litigation has centered on the construction permit, historic preservation, endangered species, and impacts on aircraft.[77] Indeed, the Cape Wind project is likely the most litigated wind project in the world. Given that the project also is an aberration given its unique geography (situated in a sound, in federal waters that are surrounded on all sides by state waters and islands resulting in social challenges and more conflict with, e.g., aircraft, comparatively close to shore (10 km), and in one of the few areas of federal water in which a state manages federal fish resources, and with an opposition supported by powerful financial and political interests), not too much should be made of the litany of litigation as being suggestive of more general trends; still it is useful to consider the EIS litigation.[78]

In *Public Employees for Environmental Responsibility v. Hopper*, 827 F.3d 1077 (D.C. Cir. 2016), among other things, the Plaintiffs contended that the EIS inadequately assessed the seafloor and subsurface hazards. The DC Circuit Court of Appeals agreed, holding that BOEM had failed to take a "hard look" at the geophysical and geological environment and that it could not rely on additional surveys to be undertaken after leasing but prior to construction, as "NEPA does not allow agencies to slice and dice proposals in this way."[79] Indeed, the EIS must "look beyond the decision to lease and consider predictable consequences of that decision."[80] Rather than vacate the lease, however, the court merely prohibited Cape Wind from engaging in construction until adequate geologic investigations are completed. Nevertheless, the decision suggests that future decisions to lease will likely be delayed until some geophysical and geotechnical work is undertaken at a site. Paradoxically, the Court found BOEM appropriately granted Cape Wind a departure from the normal course under its OCSLA regulations, when it permitted Cape Wind to defer undertaking certain geologic studies prior to being issued a construction and operations permit, an action that falls well after lease issuance.[81]

77. *See* e.g., Litigation History of Cape Wind (in chronological order), http://www.capewind.org/sites/default/files/downloads/Litigation%20History%20of%20Cape%20Wind%20May%202%202014.pdf (last visited July 26, 2016). Note that this is compiled by the developer and has not been updated since 2014, and thus, does not include the 2016 of the US Court of Appeals for the DC Circuit.
78. Prices for offshore wind power in power purchase agreements are also not without controversy. *See* "Petition of Massachusetts Electric Company and Nantucket Electric Company, each d/b/a National Grid, for approval by the Department of Public Utilities of two long-term contracts to purchase wind power and renewable energy certificates", available at http://archives.lib.state.ma.us/bitstream/handle/2452/54309/ocn690297359.pdf?sequence=4&isAllowed=y (last visited July 26, 2016) and In re Review of Proposed Town of New Shoreham, 25 A.2d 482 (S Ct RI 2011).
79. *Public Employees for Environmental Responsibility v. Hopper*, https://www.cadc.uscourts.gov/internet/opinions.nsf/9E052DE574BD496C85257FE700503890/$file/14-5301-1622978.pdf, Slip op. at 9 (last visited January 3, 2017).
80. *Id.*
81. *Id.* at 10–11.

[C] Planning the Legal Regime: Marine Planning and/or Land Planning

In 2010, President Obama by Executive Order,[82] set out a Nation Ocean Policy that seeks to "ensure the protection, maintenance, and restoration of the health of ocean, coastal and Great Lakes ecosystems and resources," established the National Ocean Council to provide national guidance on the federal agencies' implementation of the order, and sought to have the federal agencies work in a more integrated fashion amongst themselves and with states, tribes and ocean and coastal stakeholders. The order also provided for the creation of regional planning bodies that would devise regional ocean and coastal spatial plans, however, participation is voluntary and the decision of whether to create a given regional body is up to the states in a given region.[83] Two regions, the Northeast, which is comprised of the New England states, and the Mid-Atlantic, which runs from Virginia to New York, chose to create regional plans. In each region, final plans were adopted in late 2016.[84]

These plans, which are based on the President's Executive Order authority rather than on Congressional legislation, seek to improve coordination, consultation and governance rather than impose new regulatory requirements. In that regard, they are different than the land management plans of US federal land management agencies such as the Bureau of Land Management (BLM)[85] and the National Forest Service (USFS),[86] which were both established by statute.[87]

The process of creating the marine spatial plans has provided an important avenue for public engagement, consultation and input on ocean and coastal issues, including offshore wind power. That said, the effect of these marine spatial plans on offshore wind power is at this time uncertain. Yet, we do know that without such data and planning in the past, BOEM did designate offshore wind power lease areas without full consultation of other stakeholders leading to conflicts, for example, between designated shipping lanes and BOEM-identified proposed leasing sites off the Maryland and Massachusetts coasts and between offshore wind power and fishing interests in Massachusetts and New York, with the latter leading to litigation that commenced in late 2016 immediately preceding the leasing of the New York Wind Energy Area.[88]

---

82. Executive Order 13547 –Stewardship of the Ocean, Our Coasts, and the Great Lakes (July 19, 2010), available at https://www.whitehouse.gov/the-press-office/executive-order-stewardship-ocean-our-coasts-and-great-lakes (last visited July 26, 2016).
83. National Ocean Council, Marine Planning, https://www.whitehouse.gov/administration/eop/oceans/marine-planning (last visited July 26, 2016).
84. *See* http://www.boem.gov/Mid-Atlantic-Regional-Planning-Body/ (Mid-Atlantic Planning Body) (last visited January 3, 2017); https://www.boem.gov/Ocean-Action-Plan/ (Mid-Atlantic Plan) (last visited January 3, 2017) http://midatlanticocean.org/data-portal/ (Mid-Atlantic Data Portal) (last visited January 3, 2017); http://neoceanplanning.org/ (Northeast Planning Body) (last visited January 3, 2017); http://neoceanplanning.org/plan/ (Northeast Plan) (last visited January 3, 2017); http://www.northeastoceandata.org/ (Northeast Data Portal) (last visited January 3, 2017).
85. *See* Federal Land Policy and Management Act of 1976 (FLPMA), 43 USC §§ 1701 et seq.
86. *See* National Forest Management Act of 1976 (NFMA), 16 USC §§ 1600 et seq.
87. Firestone, *supra* n. 17.
88. *See* http://atlanticscallops.org/ (last visited January 3, 2017).

Coordination and consultation will help to de-conflict similar future situations, which given the present uncertainty over the financial prospects of offshore wind power development in the US should prove beneficial. As well, potential conflicts over submerged cultural resources and with Indian tribes would seem to diminish with better data, communication and appreciation of the interests at stake. Indeed, the US has witnessed some of the fruits of ocean planning for offshore wind power in Rhode Island. While Rhode Island undertook marine spatial planning on a micro-basis,[89] and the offshore wind power project at issue – Deepwater Wind's Block Island project – is only six turbines totaling 30 MW, it is importantly the only project that the Americans may see completed prior to 2020, providing some evidence of the benefits of marine spatial planning.[90]

[D] Public Participation Scheme

Public Participation in the US is primarily driven by NEPA. NEPA provides a role for the public in determining the proper scope of the EIS investigation[91] and in commenting on the draft EIS.[92] These are, however, not opportunities for active engagement between citizens and government. Rather, they take the form of notice and the opportunity for interested parties to provide written or oral comment, with the government's response to comments coming only later and in writing. In addition, BOEM has been pro-active in engaging state government stakeholders prior to designation of wind energy areas (WEAs). Maryland provides a good example, where BOEM held six stakeholder meetings between 2010 and 2013, a public seminar on the Maryland WEA in 2014, and a meeting on fishing best management practices and mitigation measures in 2013.[93]

[E] Grid

The US offshore wind power regulatory regime is flexible in regard to offshore transmission. Offshore wind power developers can gain approval to lay both inter-array cables and collection cables that run to shore in conjunction with their lease[94]

89. *See* RI Ocean SAMP, http://seagrant.gso.uri.edu/oceansamp/. Rhode Island undertook planning under the authority of the CZMA, 16 USC § 1452, which encourages the creation of "special area management plans." Massachusetts also has been a leader in marine spatial planning. *See* http://www.mass.gov/eea/waste-mgnt-recycling/coasts-and-oceans/mass-ocean-plan/.
90. For an evaluation of some of the critical aspects of a state regulatory regime, *see* Amardeep Dhanju and Jeremy Firestone, Access System Framework for Regulating Offshore Wind Power in State Waters, *Coastal Management* 37(5): 441–478 (2009).
91. http://www.boem.gov/Environmental-Stewardship/Environmental-Assessment/NEPA/policy/eis/process.aspx (last visited July 27, 2016).
92. 40 CFR 1503.
93. *See* http://www.boem.gov/Maryland/ (last visited July 26, 2016); and http://www.boem.gov/Fishing-Offshore-Wind-Mitigation-Measures-Development-Workshops/ (last visited July 26, 2016).
94. 30 CFR § 585.200.

while other entrepreneurs can seek rights-of-way[95] to construct standalone transmission, such as Google/Trans-Elect did when it sought to build an offshore wind backbone transmission line in the mid-Atlantic.[96] Irrespective of the legal authority, those laying cables have to be cognizant of existing cables buried beneath the seafloor, which are primarily telecommunications in orientation, while others have national security implications.[97] However, the larger concerns may be those associated with cables coming ashore[98] - where offshore wind comes home as far as coastal residents are concerned - and the effect on submerged cultural resources.[99]

[F] Other Concerns and Legal Regime

A number of other concerns - that is, conflicts, have arisen in the offshore wind context in the US, including those related to national security, ship safety, commercial fishing, endangered species and aesthetics, which bear mention. It should be noted that many of these conflicts have arisen because BOEM was undertaking pre-designation activities and/or designating leasing areas for offshore wind power development prior to the time that regional marine spatial planning came to the fore in either the northeast or mid-Atlantic and before BOEM had fully established more collegial relationships with its counterparts in other federal agencies as an outgrowth of those regional planning processes. As a result, we might expect the future to be less conflictual than the past.

On the national security front, there have been concerns more generally about the effect of wind turbines on military radar systems, while more specific concerns have arisen regarding conflicts with naval training and testing areas, particularly off the coasts of Virginia and North Carolina. More widespread conflicts have resulted with marine transportation and commercial fishing interests. Although some marine transportation conflicts have arisen in regard to large commercial ships (e.g., container ships and oil tankers) and the USCG has taken an aggressive position regarding

95. 30 CFR §§ 585.300–585.304.
96. *See* www.atlanticwindconnection.com (last visited July 26, 2016). Delays in the development of offshore wind resulted in retrenchment of the proposal geographically and re-alignment of its purpose from facilitating offshore wind to one being justified based on its ability to facilitate the movement of energy to markets that have higher prices for energy.
97. *See* Naval Sea Floor Cable Protection Office, http://www.navfac.navy.mil/products_and_services/ci/products_and_services/naval_ocean_facilities_program/sea_floor_cable_protection_nscpo.html.
98. Bethany rejects BlueWater Wind Proposal, Coastal Point (August 26, 2011), available at http://www.coastalpoint.com/content/bethany_rejects_bluewater_wind_proposal (last visited July 26, 2016); Narragansett Town Council spars with Deepwater Wind, The Block Island Times (July 2, 2013), available at http://www.blockislandtimes.com/article/narragansett-town-council-spars-deepwater-wind/33405 (last visited July 26, 2016).
99. *See Narragansett Indian Tribe v. Narragansett Electric Company dba National Grid*, CA 16-216, Complaint for Declaratory and Injunctive Relief, May 13, 2016) available at https://turtletalk.files.wordpress.com/2016/05/1-complaint.pdf (last visited July 26, 2016); *See also* Judge denies tribe's request to suspend construction of cable from Narragansett to Block Island for wind farm, The Bulletin (May 17, 2016) http://www.norwichbulletin.com/article/20160517/NEWS/160519552 (last visited July 26, 2016).

routing,[100] common sense solutions[101] exist regarding those vessels. More difficult may be tugs and barges,[102] which by their nature transit close to the coast in calmer waters and in areas separate from those larger vessels. Conflicts with commercial fishers also have arisen in Maryland, Massachusetts and New York[103] and likely to arise in the northern half of the New Jersey Wind Energy Area because of the presence of sessile (immobile) species (clams).

On the environmental front, issues regarding great whales, and in particular, the critically endangered north Atlantic Right Whale (NARW), are likely to animate conversations and decisions regarding acceptable locations and acceptable construction practices. This is particularly so given that, while Europe has marine mammals, it has a dearth of great whales, and thus, the twenty-five years of European offshore wind experience tells us little of how, for example, turbine construction and operation may have affect whale migratory behavior. Moreover, given the recent expansion of right whale critical habitat,[104] and a recent decision out of the 9th Circuit Court of Appeals[105] on the need to ensure the "least practical adverse impact" on marine mammals when evaluating the terms of an incidental take permit, even in those instances when population levels are "not threatened significantly," developers are likely to experience increased difficulty and delay in obtaining such permits.

Lastly, aesthetic issues can be prominent given the effect that offshore wind turbines can have on day and night time (given lights to assist air and sea navigation) views,[106] including effects on individuals' attachment to particular places and their sense of place. Indeed as Kempton et al., find, at least for some, "there appears to be something special about the ocean."[107] As far as regulatory implications of the "view," such effects are primarily evaluated under NEPA; however, the issue of aesthetics may be most germane when offshore wind projects are proposed to be located adjacent to features administered by the National Park Service (NPS) - features which together comprise a system whose fundamental purpose includes "scenery" conservation.[108]

100. *See* USCG, Atlantic Coast Port Access Route Study (AC PARS), available at http://www.uscg.mil/lantarea/acpars/ (last visited July 26, 2016).
101. Kateryna Samoteskul, Jeremy Firestone James J. Corbett, John Callahan, Analysis of Vessel Rerouting Scenarios to Open Areas for Offshore Wind Power Development Reveals Significant Societal Benefits, *Journal of Environmental Management*, 141: 146–154 (2014).
102. BOEM, Announcement of Area Identification - Commercial Wind Energy Leasing on the Outer Continental Shelf Offshore North Carolina (2014), available at http://www.boem.gov/NC_AreaID_Announcement/ (last visited July 26, 2016).
103. *See supra* n. 87.
104. 81 Fed Reg 4838-4874 (January 27, 2016) to be codified at 50 CFR § 226.203.
105. *NRDC v. Pritzker*.
106. Patrick Devine-Wright. Rethinking NIMBYism: The role of place attachment and place identity in explaining place-protective action, Journal of Community and Applied Social Psychology, 19: 426–441 (2009); Andrew Krueger, George Parsons, and Jeremy Firestone, Preferences for Offshore Wind Power Development: A Choice Experiment Approach, *Land Economics*, 87(2): 268–283 (2011); Alison Bates, and Jeremy Firestone, A Comparative Assessment of Offshore Wind Power Demonstration Projects in the United States, *Energy Research and Social Science*, 10: 192–205 (2015).
107. Willett Kempton, Jeremy Firestone, Jonathan Lilley, Tracy Rouleau, and Phillip Whitaker, The Offshore Wind Power Debate: Views from Cape Cod, *Coastal Management*, 33(2): 121–151, 132 (2005).
108. 54 U.S.C. § 100101.

The NPS' mandate recently played out off the coast of North Carolina, with BOEM ultimately acquiescing to an NPS request that a wind energy area not be located within 33.7 nautical miles of the Bodie Island Lighthouse.[109]

§6.04 CHALLENGES AND SOLUTIONS[110]

[A] Institutional Design: Challenges and Solutions

Unlike offshore oil and gas, which is sold in regional markets that are dramatically influenced by global prices, electricity sales are much more closely linked to individual state markets. In the present institutional design, however, state electricity markets play little to no role in upfront selection of developers. Instead, sites are auctioned to the highest bidder without regard to bidders relative competency and financial wherewithal to complete these ambitious projects so long as the bidder passes minimum standards. This leads to competition over upfront bonus payments to the federal government rather than having developers vetted at the state level by electric utility commissions, where the focus would be on price for the product delivered to consumers (megawatt-hours (MWh)) and the competency and experience of the developer. The solution – to change the balance of power and the order of tasks between the landlord (the federal government) and the purchaser of services (the states) – is straightforward, but may require legislative action, which is not an easy task in the present US political environment, but one that is worthy of pursuit.

[B] Incentives: Challenges and Solutions

Offshore wind power present a number of financial challenges, including a regulatory regime that generates very long development time horizons, which require investors to wait many years to see a return on investment; large capital costs that require assurances in the form of PPAs or OREC laws in order for the projects to be financeable; and prices for the initial set of projects that are likely to be quite high until economies of scale are realized and local infrastructure, local manufacturing of turbine components and the supply chain matures. The US offshore wind power sector is further burdened by low natural gas prices, ironically by inexpensive and bountiful land-based wind resources, and by the failure to price the external costs of energy production (climate, health and ecological).

While the US will likely not address all of these challenges, a number of solutions could be implemented now, including the adoption of the Carper-Collins ITC bill and state offshore wind power specific RPS laws or set asides. As well, the expansion of RGGI and the California Cap and Trade law to other jurisdictions and the imposition

109. http://www.boem.gov/NC_AreaID_Announcement/ (last visited July 27, 2016).

110. For more detail on recommendations for the next ten years, *see* generally, Jeremy Firestone, Cristina A. Archer, Meryl P. Gardner, John A. Madsen, Ajay K. Prasad, Dana E. Veron, The time has come for offshore wind power in the US, *Proceedings of the National Academy of Sciences*, 112(39): 11985–11988, doi: 10.1073/pnas. 1515376112 (2015).

within the RGGI states and California of much tighter caps to increase the price of carbon allowances to a figure closer to social cost of carbon[111] would be welcomed.

[C] Regulations: Challenges and Solutions

Finally, the present offshore wind power regulatory regime in the US is based too closely on the regime for offshore oil and gas, which poses its own unique catastrophic risks that must be guarded against - risks that are not present in the offshore wind power context. Moreover, any risks imposed by offshore wind must be balanced against the risks of delay (or failure to develop at all) to climate and to health, given that energy generated from offshore wind turbines is most likely to displace coal and natural gas generation.[112] The existing bulky regulatory regime leads not only to delayed development, but to more expensive development as well. The solution to this problem is straightforward - regulatory reform, which the Trump Administration will hopefully advance as part of a broader strategy to make offshore wind its own.

§6.05 CONCLUSION

Commercial-scale offshore wind power development in the United States faces an uncertain future. The Obama Administration laid the bedrock for future administrations by leasing large tracts of ocean expanse in the Atlantic and making significant strides in the Pacific. But a federal offshore wind power infrastructure strategy along with the regional cooperation necessary to make this capital-intensive endeavor more financially manageable and an institutional design that relies on more state-centric competition over price to residential ratepayers rather than bonus money to the federal treasury has been missing. Despite these deficits, states like Massachusetts, Maryland and New York are persevering and other states such as Hawaii and California show promise; it may very well be that offshore wind power grows from the bottom up in the US. The fact that companies such as DONG Energy, the largest offshore wind developer in Europe and the oil and gas producer Statoil are playing in the US market suggests that it might be prudent to bet on those states.

---

111. Interagency Working Group on the Social Cost of Greenhouse Gases, United States Government, Technical Support Document: Technical Update of the Social Cost of Carbon for Regulatory Impact Analysis - Under Executive Order 12866 (Revised August 2016), available at https://www.epa.gov/sites/production/files/2016-12/documents/sc_co2_tsd_august_2016.pdf (last visited January 3, 2017). *See also* https://www.epa.gov/climatechange/social-cost-carbon (last visited January 3, 2017).
112. *See* e.g., GE Energy Consulting, PJM Renewable Energy Integration Study, Executive Summary Report (2014), available at https://www.pjm.com/~/media/committees-groups/committees/mic/20140303/20140303-pris-executive-summary.ashx (last visited July 28, 2016).

CHAPTER 7

Legal Framework to Develop Offshore Wind Power in Australia

Alex Wawryk

§7.01 INTRODUCTION

Australia is a party to both the United Nations Framework Convention on Climate Change (UNFCCC) and the Kyoto Protocol.[1] In 2010 Australia agreed, under the Cancún Agreements to the UNFCCC, to reduce its emissions of greenhouse gases to 5% below 2000 levels, by 2020. More recently, Australia's long-term target, expressed in its Nationally Determined Contribution (NDC) under the Paris Agreement,[2] is to reduce GHG emissions to 26%-28% below 2005 levels by 2030. Australia also accepted an international obligation under the Doha Amendment to the Kyoto Protocol for the second period of the Protocol, although Australia has not yet ratified the Doha Amendment.[3] Australia's second-period Kyoto Protocol commitment is an unconditional target, reflecting its commitment under the Cancún Agreements: by 2020, to reduce GHG emissions to 5% below 2000 levels.[4]

1. United Nations Framework Convention on Climate Change, Adopted June 14, 1992, UNCED Doc A/CONF.151/5/Rev.1, (Vol. I), Annex I, June 13, 1992, 31 *ILM* 874 (1992); Kyoto Protocol to the United Nations Framework Convention on Climate Change, Report of the Conference of the Parties at its Third Session, December 1-11, 1997, U.N. Doc FCCC/CP/1997/7/Add.1, March 18, 1998, Annex.
2. United Nations Framework Convention on Climate Change, Report of the Conference of the Parties at its Twenty-First Session, November 30-December 11, 2015, FCCC/CP/2015/L.9/Rev.1, December 12, 2015, Annex.
3. Conference of the Meeting of the Parties, Decision 1/CMP.8, FCCC/KP/CMP/2012/13/Add.1, *Amendment to the Kyoto Protocol pursuant to its Article 3, paragraph 9 (the Doha Amendment).*
4. Australia also committed to a second, conditional target under the Doha Amendment, which was largely forgotten by the federal Liberal government: to reduce GHG emissions by up to either 15% or 25% by 2020, depending on whether there is a global agreement to secure atmospheric

To help meet its international greenhouse gas emission reduction obligations, the Australian government established measures to encourage the generation of electricity from renewable energy, most notably the Renewable Energy Target (RET). Underpinned by the RET, onshore wind energy has been the fastest growing renewable energy source in Australia in recent years. However, there is as yet no offshore wind energy in Australia despite significant offshore wind resources, largely because it is uncompetitive in terms of cost. Offshore wind energy is more expensive than onshore wind farms.[5] The cost of foundations and turbines, installation, maintenance and repair, and decommissioning are all higher in the marine environment, and tend to be more expensive the deeper the water and the further the distance from shore. In many coastal areas in Australia, the continental shelf falls steeply, making the water deep and correspondingly offshore wind energy more expensive, with floating turbines the only feasible option.[6]

With its massive cheap resources of coal, significant gas resources, and large areas of land for renewable energy such as wind and solar onshore, there has been little incentive to date to develop offshore wind energy in Australia. It has seemed unlikely that wind developers will look offshore in the near future,[7] and supporting the development of an offshore wind energy has not been a priority of any of the Australian governments.

However, offshore wind energy has certain advantages, one of which is that average wind speeds offshore are higher and more consistent, leading to greater and more reliable capacity offshore.[8] Offshore wind farms may use larger turbines and more of them because of a lack of community objection to aesthetics, thus the higher capacity of offshore farms may offset the high costs of installation, maintenance and repair.[9] Also, the cost of compensation or rent for access to private land onshore, calculated over a twenty-year period, can add a significant cost burden to wind energy proponents that need not be paid for offshore developments, and help to equalize the cost of development offshore.

---

stabilization at 450ppm CO2-e and whether major developing economies commit to substantially restraining their emissions and advanced economies take on commitments comparable to Australia's.

5. P Breeze, *Power Generation Technologies* (2nd ed, Elsevier, 2014) 223, 233–234; J Kaldellis and M Kapsali, "Shifting Towards Offshore Wind Energy – Recent Activity and Future Development" (2013) 53 *Energy Policy* 136, 142–144; D Jeng and Y Zheng, "Energy from Offshore Wind: An Overview" in Gavin Birch (ed) *Water Wind Art and Debate: How Environmental Concerns Impact on Disciplinary Research* (University of Sydney Press, 2007) 244, 251–252.
6. Floating turbines are the only feasible option in waters of greater than 60m depth: European Wind Energy Association, *Deep Water: The Next Step for Offshore Wind Energy* (July 2013) 12, http://www.ewea.org/publications/reports/deep-water/, accessed October 19, 2015.
7. *See* for example, L Carson, "Wind Energy", *Australian Energy Resource Assessment – Second Edition* (Geoscience Australia, Canberra, 2014) 239, 253.
8. P Breeze, above n. 5, 233; J Kaldellis and M Kapsali, above n. 5, 141.
9. M Boelen, I Bishop and C Petit, "Selecting Offshore Renewable Energy Futures for Victoria" (2010) 38 *The International Archives of the Photogrammetry, Remote Sensing and Spatial Information Sciences* 478, 478; D Jeng and Y Zheng, above n. 5, 251.

Technological development should bring down costs in the future,[10] while cost barriers can also be addressed by ongoing government support for renewables such as the RET. Some studies have suggested that costs of offshore wind energy are competitive under certain conditions and in certain places off the Australian coast,[11] and have identified a selection of the most likely sites for offshore wind industry in Australia.[12] Potential sites include area off the Western Australian coast near the South-West Integrated System (SWIS) electricity grid; and near the cities of Whyalla (South Australia, (SA)), Gladstone (Queensland), Rockhampton (Queensland), Bundaberg (Queensland), Mackay (Queensland) and Melbourne (Victoria).

With increasing restrictions on the availability of land onshore in some States because of community objections to wind farms on the grounds of aesthetics, noise and even health concerns, it is becoming more likely that an offshore wind energy industry may emerge in Australia. Recently, at least one wind energy developer has been engaged in government discussions regarding the establishment of a large offshore wind farm. Furthermore, in November 2015, federal Environment Minister Greg Hunt flagged a push towards investment "on a grand scale" in offshore wind in Australia, although on the proviso that it is cost-competitive.[13]

This chapter examines the legal regime for the development of offshore wind energy in Australia. Section §7.02 provides a very general overview of the institutional design features for offshore wind energy, providing a snapshot of the many and various state and federal authorities that would be involved in the development of an offshore wind energy facility. Section §7.03 describes the legal regime for offshore wind energy in Australia. It addresses three major issues. First, Section §7.03 describes the major Commonwealth laws and policies that provide incentives for the development of renewable energy, including offshore wind energy, in Australia. Second, section §7.03 analyzes and critically assesses the licensing regime for offshore wind energy, focusing on the South Australian legal regime as an example of State/Territory law and regulation, and the relevant law applicable in Commonwealth waters. Third, section §7.03 contains a very brief introduction to and critique of the National Electricity

10. Citing other studies, the European Wind Energy Association has stated that "In an optimum regulatory and competitive market", "bigger turbines utilising cutting edge technology will increase yields and cut costs by as much as 17% by 2020" and "a 39% reduction in costs could be achieved by 2023": EWEA, http://www.ewea.org/policy-issues/offshore/, accessed September 23, 2015. *See also* The Crown Estate, *Offshore Wind Cost Reduction Pathways Study* (May 2012), http://www.thecrownestate.co.uk/energy-and-infrastructure/offshore-wind-energy/working-with-us/strategic-workstreams/cost-reduction-study/, accessed October 19, 2015.
11. S Christos, *Investigation of the Potential to Implement Offshore Wind Energy Technology in Victoria*, Australia, Masters Thesis in Sustainable Development, Department of Earth Sciences, Uppsala University, 2015; M Batchelor, *Feasibility of Offshore Wind in Australia*, Thesis submitted to Murdoch University, November 2012.
12. E Messali and M Diesendorf, "Potential Sites for Off-Shore Wind Power in Australia" (2009) 33 *Wind Engineering* 335; M Boelen, I Bishop and C Petit, above n. 9; D Jeng and Y Zheng, above n. 5.
13. G Parkinson, "Hunt Flags Move to Offshore Wind on 'Grand Scale'. Is He Serious?" *RenewEconomy*, November 4, 2015, http://reneweconomy.com.au/2015/hunt-flags-move-to-offshore-wind-on-grand-scale-is-he-serious-64652 (accessed June 10, 2016).

Market (NEM) rules in Australia. Section §7.04 discusses challenges for the development of the offshore wind energy industry in Australia, and possible solutions, while section §7.05 concludes.

§7.02 INSTITUTIONAL DESIGN FOR OFFSHORE WIND

Australia has a federal system of government, comprised of the Commonwealth government, and the governments of the six Australian States and two Territories.[14] The issue of which government has the right and responsibility to pass legislation for the assessment and approval of offshore developments is a matter of constitutional law,[15] Commonwealth and state law,[16] and political agreement.[17] The State and Northern Territory governments regulate onshore developments, including on the coastline and in internal waters, and developments offshore in "state coastal waters" (generally speaking, waters seaward from the low water mark to 3 nautical miles (nm)).[18] The Commonwealth legislates over "offshore areas" in Commonwealth waters, that is, waters seaward from 3 nm to 200 nm (the outer limit of the EEZ), and on the continental shelf where that extends beyond the 200 nm limit of the EEZ.

[A] The "Main" Authority During the Development of Offshore Wind Power

Unlike the major existing offshore industries such as aquaculture, fisheries, minerals and petroleum, which are regulated under Commonwealth and/or State sectoral-specific legislation, no legislation specifically governs offshore renewable energy projects. A proposal for an offshore wind energy facility will be assessed and approved under general state and/or Commonwealth planning/development law. The main authority for a proposal sited in State coastal waters will be the State planning department/authority. As each State and the Northern Territory have their own legislation and regulatory authorities, there are seven separate legal jurisdictions and sets of regulatory bodies at the state/NT level. Furthermore, Commonwealth legislation may also apply to a development in state coastal waters, adding another legal jurisdiction and more regulatory authorities to the mix.

The situation for a proposal sited in Commonwealth waters is less clear. As will be seen in the discussion below regarding Licensing Regimes, there is no general

14. The six states are New South Wales, Queensland, South Australia, Tasmania, Victoria and Western Australia. The two Territories are the Northern Territory and the Australian Capital Territory. The ACT is landlocked.
15. The Commonwealth Constitution; *New South Wales v. Commonwealth* (1975) 135 CLR 337.
16. *Seas and Submerged Lands Act 1973* (Cth); *Coastal Waters (State Powers) Act 1980* (Cth); *Coastal Waters (State Title) Act 1980* (Cth).
17. *Offshore Constitutional Settlement: A Milestone in Co-operative Federalism* (AGPS, Canberra, 1980) and *Offshore Constitutional Settlement: Selected statements and documents 1978-79* (Commonwealth of Australia, 1980), <http://www.ret.gov.au/resources/Documents/upstream-petroleum/Offshore_Constitutional_Settlement.pdf>, July 31, 2013.
18. The Australian Capital Territory is landlocked.

planning framework in Commonwealth waters to govern offshore wind development in Commonwealth waters, nor are there arrangements to deal with a proposal where turbines span State and Commonwealth waters. It is not clear which government authority would preside over a wind farm situated in Commonwealth waters, with cables running through state coastal waters to shore, and substations constructed on state land. While the Commonwealth Department of Energy and the Environment is the agency responsible for assessing environmental impacts and the Environment Minister for granting environmental approvals under the Commonwealth *Environment Protection and Biodiversity Conservation Act 1999* (Cth) ("EPBC Act"), an environmental approval is not equivalent to a license granting the holder certain legal rights to exploit the resource, unlike, for example, petroleum authorizations under the *Offshore Petroleum and Greenhouse Gas Storage Act 2006* (Cth).

[B] The "Supplementary" Authorities During the Development of Offshore Wind Power

A plethora of government agencies at the federal and state level are likely to be involved in the general development of an offshore wind industry, and/or more particularly, the assessment and approval of a particular offshore wind energy farm. In some cases, state planning legislation will require referral to a particular government agency or Minister, usually the Environment Minister, but even without referral, it would be expected (as a minimum) that the agencies listed in the tables below would be consulted and/or put in submissions given the impact in their area of regulation management. These tables summarize key government agencies; other community and representative bodies will also participate in the development process, for example, Native Title representative bodies, representing the interests of Aboriginal native title holders; environmental community groups; and the fishing industry, to name but a few.

[1] Central Government Level

Regulation of the RET	Clean Energy Regulator
Provision of incentives	Australian Renewable Energy Agency Clean Energy Finance Corporation
National Electricity Market – rules and regulation	Australian Energy Market Commission (AEMC); Australian Energy Regulator (AER); Australian Energy Market Operator (AEMO)
Environment	Department of Energy and the Environment
Review of climate policy	Climate Change Authority
Maritime safety and navigation	Australian Maritime Safety Authority
Fishing	Department of Agriculture and Water Resources

Offshore Petroleum	Department of Industry, innovation and Science; the Joint Authorities for the offshore area of each state and the Northern Territory, comprising the responsible Commonwealth Minister and the relevant state and NT Resources Minister; the National Offshore Petroleum Safety and Environment Management Authority; National Offshore Petroleum Titles Administrator
Defence	Department of Defence

[2] State Government Level

Electricity generation and transmission licenses, cables	State/Territory electricity industry regulators e.g., in SA, the Essential Services Commission of SA
Tenure	Relevant State department e.g., in SA, the Department of Transport
Environment	Range of authorities e.g., in SA, the Department of Environment, Water and Natural Resources; Environment Protection Authority; Coast Protection Board; Native Vegetation Council; South Australian Heritage Council
Protection of Aboriginal heritage	Relevant State department e.g., in SA, the Aboriginal Affairs and Reconciliation Division of the Department of State Development
Aquaculture and Fishing	Relevant State department e.g., in SA, Primary Industries and Regions SA
Offshore Petroleum	Relevant State department e.g., in SA, the Department of State Development
Defence	Commonwealth Department of Defence
Tourism	South Australian Tourism Commission
Expansion of Ports	Various Ports authorities

[3] Local Government

Although local governments assess and approve developments under certain planning law procedures in the Australian states, the planning procedures most likely to apply to a complex and expensive infrastructure project such as offshore wind energy will limit the role of local government. While local councils and agencies will participate through the consultation processes, they will usually not be responsible for assessing or approving large offshore wind developments. These would be expected to be "called in" and assessed by a state agency/Minister under a major development or significant projects procedure. An example of this type of procedure under South Australian planning law is described under the heading "Licensing Regime," below.

§7.03 LEGAL DESIGN FOR OFFSHORE WIND

[A] Incentives

Currently there is no specific government policy, law or measures to encourage offshore wind energy, nor is it likely there will be any specific policy support for offshore wind energy in the near future. Any government support received by offshore wind energy will stem from the general schemes for encouraging renewable energy established by the federal Australian government, or various state governments. As it is beyond the scope of this chapter to examine all the state government initiatives to reduce greenhouse gas emissions, this section will focus on the major measures established by the federal government to support renewable energy development, including offshore wind energy.

[1] National Market-Based Mechanisms to Decrease GHG Emissions and/or Increase the Uptake of Renewable Energy

[a] Emissions Trading

By putting a price on GHG emissions, emissions trading schemes make GHG-emitting fuels more expensive, thereby improving the cost competitiveness of alternative energy sources such as offshore wind energy. Emissions trading has been a controversial issue in Australia since the 1980s. The first national Australian cap-and-trade emissions trading scheme, the Carbon Pricing Mechanism (CPM), was established by the federal Labor Government under the Prime Ministership of Julia Gillard in 2011 through the *Clean Energy Act 2011* (Cth). The CPM was designed to operate with a fixed price per tonne of CO_2 equivalent greenhouse gases from 2012–2015. From July 1, 2015, there was to be an economy-wide cap on the emission of greenhouse gases consistent with Australia's international obligations to reduce GHG emissions, with a flexible price.

Under the CPM, liable entities had to buy, or be allocated with Australian Carbon Credit Units (ACCUs) and relinquish one carbon unit for each tonne of carbon pollution produced each year through the surrender of "eligible emissions units" to the regulator. Liability entities who failed to surrender sufficient units had to pay a "unit shortfall charge" in the fixed price period of 1.3 times the fixed price per unit. In the flexible price period, this was to be double the annual average price of units. Liable entities with obligations under the scheme were first, a person with operational control of a facility that emitted over 25,000 tonnes of "scope 1 emissions" of CO2-e GHGs annually; and second, retail suppliers of natural gas. About 500 of the biggest polluters were liable under the CPM. The scheme applied to carbon dioxide, methane, nitrous oxide and perfluorocarbons. Sectors covered by this scheme included: stationary energy; industrial processes; fugitive emissions; and emissions from non-legacy waste. Rail, domestic aviation and shipping transport fuels were covered by an equivalent carbon price through separate legislation. Various assistance packages were offered to "emissions-intensive trade-exposed" and other affected industries and households were given tax cuts.

During the fixed price period, "eligible emissions units" included: carbon units purchased from the Australian Government at the fixed price, or obtained for free as part of an allocation under an assistance program; and, to a limit of 5% of its total liability, ACCUs issued under the Carbon Farming Initiative, a separate scheme aimed at encouraging carbon offsets through sequestration and emissions avoidance activities in the agricultural/land use sector under the *Carbon Credits (Farming Initiative) Act 2011* (Cth). In the flexible price period, "eligible emissions units" were to include: carbon units purchased from the Australian Government at auction, or obtained through free allocation under assistance programs, or purchased under the secondary market for permits; ACCUs obtained from the Carbon Farming Initiative; and, with a 50% limitation until 2020, international emissions units created under the Kyoto Protocol or under recognized regional/national schemes such as the European and New Zealand emissions trading schemes. A treaty for a two-way link with the European Union emissions trading scheme was to have been agreed by July 1, 2015 and implemented no later than July 1, 2018.

In September 2013, a new federal Coalition government under Prime Minister Tony Abbott won office, abolishing the CPM in 2014. However, in order to get its alternative to the CPM – the "Direct Action Plan" – through the Senate, the Coalition government was forced to agree to an inquiry into an emissions trading scheme. Thus, in December 2014, the government requested the Climate Change Authority to conduct a Special Review into Australia's policies and future targets for reducing greenhouse gas emissions, in the context of its international commitments and the action of other countries.[19] The terms of reference for the Review include whether Australia should have an emission trading scheme and the conditions under which such a scheme should be established.

The final report of this special Review was due by June 30, 2016, but delayed because of the federal election on July 2, 2016. Released at the end of August 2016, the Climate Change Authority's Report did not recommend the reintroduction of a cap-and-trade emissions scheme, but rather recommended the government build upon the current Emissions Reduction Fund (ERF) and Safeguard Mechanism (*see* below), as well as introduce a market-based emissions intensity scheme for the electricity supply industry, whereby a baseline for emissions intensity per kWh would be set, reducing to zero before 2050.[20] Politically, it is now extremely unlikely the returned Coalition government under the Prime Ministership of Malcom Turnbull will reintroduce an

19. Australian Government, Climate Change Authority, "Special Review", http://www.climatechangeauthority.gov.au/special-review, accessed July 26, 2016.
20. Climate Change Authority, *Towards a Climate Policy Toolkit* (Commonwealth of Australia, 2016).

emissions trading scheme, although it remains the policy of the opposition Labor Government.

[b] *Direct Action Plan: ERF and the Safeguard Mechanism*

The ERF is the centerpiece of the "Direct Action Policy" instituted by former Prime Minister Tony Abbott's Coalition Government in place of the abolished CPM. Established through amendments to the *Carbon Credits (Farming Initiative) Act 2011* (Cth) on November 24, 2014,[21] the ERF comprises "a government fund to purchase emissions reductions, and a safeguard mechanism to ensure that these reductions are not displaced by a significant rise in emissions above business-as-usual levels elsewhere in the economy."[22]

The first element of the ERF is an AUD 2.55 billion fund for purchasing emission abatements. The ERF is a voluntary "baseline and credit" scheme. A baseline for greenhouse gas emissions for companies is set at according to a "business-as-usual" trajectory. Companies who register with the Clean Energy Regulator may seek ACCUs for emission reductions below the baseline, arising from specific "eligible offsets projects" registered with the Regulator. ACCUs can be sold to generate income, either to the federal Government through a carbon abatement contract,[23] or in the secondary market for ACCUs. There are two types of eligible offsets projects for which ACCUs may be credited. These are sequestration offsets projects, which are largely relevant to land use sectors such as agriculture and forestry; and emissions avoidance offsets projects, which are of relevance to the abatement of CO_2-e greenhouse gases from energy use.[24] Most contracts awarded to date have been for sequestration, landfill and alternative waste treatment methods, and for energy efficiency projects.

From July 1, 2016, the voluntary emissions abatement will be complemented by the ERF Safeguard Mechanism, established under Part 3H of the *National Greenhouse and Energy Reporting Act 2007* (Cth). The safeguard mechanism is designed to ensure large businesses do not increase emissions above historical levels by requiring them to keep their emissions below a set baseline.[25] Baselines will be set for individual facilities, to reflect the highest level of reported emissions for a facility over the period

21. *See also* the *Carbon Credits (Carbon Farming Initiative) Regulations 2011* and the *Carbon Credits (Carbon Farming Initiative) Rule 2015*.
22. Minister for the Environment, *National Greenhouse and Energy Reporting Act 2007, National Greenhouse and Energy Reporting (Safeguard Mechanism) Rule 2015, Draft Explanatory Statement*, 3 http://www.environment.gov.au/climate-change/emissions-reduction-fund/about/safeguard-mechanism (accessed September 3, 2015).
23. *Carbon Credits (Farming Initiative) Act 2011* (Cth) s. 20B.
24. *Carbon Credits (Farming Initiative) Act 2011* (Cth) ss. 16 and 18.
25. It has been estimated that safeguard mechanism will apply to some 261 facilities above the designated large facility threshold; that of these, 85 facilities run by 30 companies will exceed their historical baselines; and that it is unlikely any of the top twenty emitting facilities (including electricity generation facilities) will face financial penalties: Baker & McKenzie, "Australia's Emission Reduction Fund Safeguard Mechanism: Top 10 Considerations for Companies," http://bakerxchange.com/rv/ff0025da75c818a6b56f37f4274e80bd27bd7af1, March 4, 2016.

2009–2010 to 2013–2014.[26] Facilities who emit above the historical baseline must surrender credits to offset their excess emissions. ACCUs may be used to satisfy obligations under the Act.[27]

The safeguard mechanism applies only to "designated large facilities," these being businesses with direct emissions of CO_2-equivalent greenhouse gases ("scope 1 emissions") over 100,000 tonnes per financial year (the "designated large facility threshold").[28] The entity with "operational control" of a designated large facility, referred to as the "responsible emitter," will be responsible for meeting safeguard requirements.[29]

The legislation and rules set out fairly complex rules for the determination and adjustment of baselines. A special approach will be taken by the electricity sector, with one sectoral baseline applying to all "grid-connected electricity generators."[30] This includes the following five electricity grids (the "designated electricity network"): the NEM, the South West Interconnected System, the North West Interconnected System, the Darwin-Katherine Interconnected System and the Mount Isa–Cloncurry Supply Network. The sectoral baseline, which is equal to 198 Mt CO2-e greenhouse gas emissions, has been calculated as the sum of the sector's emissions in the 2009–2010 financial year. Emissions from grid-connected electricity generators are not covered as long as the sector aggregate emissions do not exceed the sectoral-baseline. If the sum of emissions exceeds the electricity sectoral baseline, grid-connected generators will then be required to keep emissions below individual facility-level baselines, set at each facility's highest annual emissions between 2009–2010 and 2013–2014.[31]

Responsible emitters must report emissions to the Clean Energy Regulator, who will issue a notice to a responsible emitter whose net emissions exceed their baseline in any given year. The responsible emitter may apply for multi-year monitoring, which allows a facility to exceed its baseline in one year if its average emissions over two or three years are below the baseline.[32] If the facility's average emissions over the

26. Detailed rules for the calculation of baselines are set out in the *National Greenhouse and Energy Reporting (Safeguard Mechanism) Rule 2015* (Cth).
27. *National Greenhouse and Energy Reporting Act 2007* (Cth), ss. 22XK-22XN.
28. *National Greenhouse and Energy Reporting Act 2007* (Cth), s. 22XJ; *National Greenhouse and Energy Reporting (Safeguard Mechanism) Rule 2015*, Rule 8.
29. *National Greenhouse and Energy Reporting Act 2007* (Cth), s. 22XH.
30. *National Greenhouse and Energy Reporting Act 2007* (Cth), s. 22XI; *National Greenhouse and Energy Reporting (Safeguard Mechanism) Rule 2015*, rr 4, 7(1)(c), 13(1)(a). *See also* Minister for the Environment, *National Greenhouse and Energy Reporting Act 2007, National Greenhouse and Energy Reporting (Safeguard Mechanism) Rule 2015, Explanatory Statement*, 19; and Australia Government, Department of the Environment, "The safeguard mechanism – Electricity sector", 2016 http://www.environment.gov.au/climate-change/emissions-reduction-fund/publications /factsheet-safeguard-mechanism-electricity-sector.
31. If the sum of emissions exceeds the electricity sectoral baseline, the Regulator must publish a statement to that effect by February 28, of the following financial year. The years up to and including the year the electricity sectoral baseline has been exceeded, and the financial year following that in which the Clean Energy Regulator publishes its statement, are "sectoral-baseline financial years". Individual grid-connected generators do not have obligations to meet individual baselines in sectoral-baseline financial years. However, after the conclusion of the sectoral-baseline financial years, grid-connected generators will be required to keep emissions below their individual facility-level baselines.
32. *National Greenhouse and Energy Reporting Act 2007* (Cth), s. 22XG.

monitoring period remain above the baseline, the Regulator has a number of enforcement options. These include: issuing infringement notices; accepting enforceable undertakings; seeking an injunction; court-imposed civil penalties; and an ongoing obligation on facility operators to rectify any emissions exceedance.[33]

[c] Renewable Energy Target

The RET is a market-based mechanism established under the *Renewable Energy (Electricity) Act 2000* (Cth), the major aim of which is to encourage the additional generation of electricity from renewable energy sources (RES-E), in order to reduce emissions of greenhouse gases from the electricity sector.[34] The RET is comprised of two sub-schemes, the Large-scale Renewable Energy Target (LRET) and the Small-scale Renewable Energy Scheme (SRES), both of which are administered by the Clean Energy Regulator.

The LRET creates a financial incentive for the generation of renewable energy from large-scale renewable power stations, such as wind offshore wind power. The LRET places a statutory obligation on "liable entities"[35] – mainly electricity retailers and wholesale electricity purchasers – to purchase and surrender large-scale generation certificates (LGCs) in order to meet the RET for the year. Liable entities are required to meet the target in proportion to the amount of wholesale electricity they purchase in a year.[36] If a liable entity submits an insufficient number of RECs to meet their annual liability, they must pay a shortfall charge.

The Clean Energy Regulator issues LGCs for the generation of electricity. One LGC is equivalent to 1 MWh of eligible RES-E generated above the "1997 eligible renewable power baseline."[37] This historical baseline was introduced to ensure RECs are only created for electricity generated from renewable energy sources that are new or additional to the amount of renewable energy generated before the Act came into effect. The RET and interim annual targets for 2016 onwards are as follows:[38]

33. *National Greenhouse and Energy Reporting Act 2007* (Cth), Pt 5; *National Greenhouse and Energy Reporting (Safeguard Mechanism) Rule 2015* (Cth), Pt 4.
34. *Renewable Energy (Electricity) (Charge) Act 2000* (Cth) s. 3. The first version of the scheme, the Mandatory Renewable Energy Target (MRET), was introduced in 2001. The name was changed to the Renewable Energy Target (RET), and the scheme was split into the LRET and SRES in 2009. The *Renewable Energy (Electricity) Regulations 2001* (Cth) provide more details on a number of issues, such as eligibility criteria for renewable energy sources and criteria for accreditation of power stations.
35. *Renewable Energy (Electricity) Act 2000* (Cth) s. 35.
36. The liability of an entity – or amount of renewable source electricity required under the scheme – is worked out by taking the amount of electricity acquired by the entity, less any exemptions, and multiplying this by the "renewable power percentage" for an entity, which is calculated using a formula set out in s. 39 of the Act.
37. *Renewable Energy (Electricity) Act 2000* (Cth) s. 17C.
38. *Renewable Energy (Electricity) (Charge) Act 2000* (Cth) s. 40.

Renewable Energy Target, 2016–2030	
Year	*Target (GWh)*
2016	21,431
2017	26,031
2018	28,637
2019	31,244
2020–2030	33,000

Certificates are not valid until registered by the Regulator.[39] LGCs are negotiated and paid for outside the REC Registry and transferred within the REC Registry. The price of LGCs is not fixed, but depends on market supply and demand.

While there have been criticisms of the RET, recent Reviews have found that it has been successful in promoting additional generation from renewable sources, with renewable energy generation almost doubling from 2001 to 2013.[40] Over the period 2001–2014, more than 400 renewable power stations with a total capacity of over 5,000 MW were installed, equivalent to about 10% of Australia's current grid-connected capacity. About 75% of this is onshore wind energy.[41]

[B] Grants and Finance

There are currently two main bodies established under federal legislation that are charged with assisting the development of renewable energy technologies through grants and finance schemes. These are the Australian Renewable Energy Agency (ARENA) and the Clean Energy Finance Corporation (CEFC).

ARENA was established as an independent statutory body on July 1, 2012 under the *Australian Renewable Energy Agency Act 2011* (Cth). The Act has two objectives: to improve the competitiveness of renewable energy technologies and to increase the supply of renewable energy in Australia.[42] ARENA's functions include providing financial assistance for: research into renewable energy technologies; the development, demonstration, commercialization or deployment of renewable energy technologies; and the storage and sharing of information and knowledge about renewable energy technologies. ARENA also is empowered to enter into agreements for the purpose of providing financial assistance, and to administer such agreements.[43]

When ARENA first began operation, the projects from nine existing grant funding programs supporting research and development (R&D) in renewable energy were consolidated into the Agency. ARENA had an initial a budget of AUD 2.5 billion until 2022, of which it has already invested AUD 1 billion in more than 250 initiatives, from

39. *Renewable Energy (Electricity) Act 2000* (Cth) s. 26.
40. Commonwealth of Australia, *Renewable Energy Target Scheme, Report of the Expert Panel* (2014), 14.
41. Australian Government, Climate Change Authority, *Renewable Energy Target Review: Report* (Commonwealth of Australia, 2014) 11.
42. *Australian Renewable Energy Agency Act 2011* (Cth), s. 3.
43. *Australian Renewable Energy Agency Act 2011* (Cth), s. 8.

basic R&D to demonstration and near-commercial deployment.[44] However, none of these are specifically directed to wind energy, as support for wind energy does not accord with ARENA's investment focus areas and priorities under its General Funding Strategy and Investment Plan. Because of the "widespread deployment and relative maturity" of wind energy technology, ARENA will only consider providing "selective support for new technologies or approaches that could reduce the cost of wind energy."[45] Wind energy may benefit from investment in one of ARENA's priority areas "integrating renewables and grids," which is intended to overcome barriers to the long-term uptake of renewables.

The CEFC was created under the *Clean Energy Finance Corporation Act 2012* (Cth) to leverage private sector funding for renewable energy and clean technology projects and identify and remove barriers to the financing of projects. The CEFC co-finances and invests in clean energy projects and technologies. The CEFC does not build or buy projects in its own right but may acquire equity interests in clean energy businesses or projects. It does not make grants. While there is a focus on projects and technologies at the later stage of development with a positive expected rate of return and the capacity to service and repay capital, the CEFC can also invest in projects in the demonstration stage that pass a risk profile.[46]

The CEFC has three broad financing structure models of investments: project finance, corporate loans and aggregation finance.[47] Of these, project financing is particularly suited to large-scale capital investments in energy infrastructure, such as onshore and offshore wind farms involving capital costs of over AUD 50 million. It involves funding a project on the basis that debt and equity will be paid back solely from the cash flow generated by the project.[48]

In contrast to ARENA, the CEFC has significant investments in onshore wind energy, with a wind energy investment portfolio valued at AUD 259 million in 2014–2015.[49] This may expand to include investments in offshore wind energy. On May 5, 2016, the federal government issued the *Clean Energy Finance Corporation Investment Mandate Direction 2016* (Cth), which gives effect to the creation of the AUD

44. Australian Government, Australian Renewable Energy Agency, "Governance and Funding Profile," http://arena.gov.au/about-arena/governance-and-funding-profile/ (accessed July 17, 2016). In September 2016, the federal government cut ARENA's funding by AUD 500 million leaving the Agency with some AUD 800 million to invest.
45. Australian Government, Australian Renewable Energy Agency, "Investment Focus Areas," http://arena.gov.au/funding/investment-focus-areas/, accessed July 17, 2016.
46. The CEFC has an average expected ROR of 6.1%.
47. Australian Government, Clean Energy Finance Corporation, "CEFC Financing Types" http://www.cleanenergyfinancecorp.com.au/what-we-do/cefc-financing-types.aspx (accessed July 17, 2016). A corporate loan provides funding on the basis that security is taken over all assets of the organization and debt will be repaid from the corporation's cash flow. Corporate loans are most appropriate when a company has a portfolio of smaller projects for which are easier to finance on a consolidated basis. Aggregation funding is used to catalyze large numbers of smaller projects in conjunction with commercial banks and other partners. Aggregate financing solutions are designed to overcome difficulties in accessing upfront capital for energy efficiency, low emissions technology and renewable energy projects.
48. *Ibid.*
49. Australian Government, Clean Energy Finance Corporation, "Investments," http://www.cleanenergyfinancecorp.com.au/investments.aspx (accessed July 17, 2016).

1 billion Clean Energy Innovation Fund (CEIF), to be administered jointly by the CEFC and ARENA.[50] The CEIF will provide up to AUD 100 million a year, from July 1, 2016 to June 30, 2026, for "debt and equity investment in emerging clean energy technology projects and businesses that involve technologies that have passed beyond the research and development stages but are not yet established or of sufficient maturity, size or otherwise commercially ready to attract sufficient private sector investment."[51] This includes "offshore energy."[52]

Furthermore, the *Clean Energy Finance Corporation Investment Mandate Direction 2016* specifically provides that in relation to all investments other than those made under the CEIF, the Corporation must include a focus on supporting emerging and innovative renewable energy technologies and energy efficiency technologies, such as offshore wind technologies.[53] Thus, the government has signaled policy support for the CEFC to invest in future development of the offshore wind energy industry in Australia.

§7.04 LICENSING REGIME

Developing offshore wind energy will raise the issue of managing competing uses of offshore areas. Offshore wind energy provides another competing use of the oceans and two types of conflict over the use of ocean space will need to be addressed.[54] The first of these is the user-environment conflict. The various potential environmental, cultural and social impacts of offshore wind energy must be assessed and mitigated.[55] Second, the potential impacts on other users of the sea must be assessed and managed. Because the costs of transmission infrastructure, and energy losses in transmission, increase with distance from the grid, proximity to a major energy load center is important, suggesting offshore wind energy facilities should be located near the coast.[56] However, this may lead to more conflicts with other users of the sea and coastal environment.

While marine spatial planning can assist in managing conflicts over competing uses of sea space, in Australia, marine spatial planning has largely developed to resolve the user-environment conflict through the use of marine plans and marine protected areas, and not multiple-use conflicts. While strategic environmental assessment (SEA) for offshore energy may be valuable for identifying and assessing the environment of marine areas, the competing interests in those offshore areas, and the areas where

50. ARENA will assess project proposals and make recommendation for funding to the CEFC. Although to date ARENA has primarily had a grant-based role, this will shift to financing projects on a debt and equity basis under the CEIF.
51. *Clean Energy Finance Corporation Investment Mandate Direction 2016* (Cth) subs 14(1).
52. The Minister for the Environment and the Minister for Finance, Explanatory Statement, *Clean Energy Finance Corporation Investment Mandate Direction 2016* (Cth), https://www.legislation.gov.au/Details/F2016L00714/Explanatory%20Statement/Text (accessed July 17, 2016).
53. *Clean Energy Finance Corporation Investment Mandate Direction 2016* (Cth) s. 13.
54. F Douvere and C Ehler, "International Workshop on Marine Spatial Planning, UNESCO, Paris, 8-10 November 2006: A Summary" (2007) 31 *Marine Policy* 582.
55. Department of Environment, Water, Heritage and the Arts, *EPBC Act Policy Statement 2.3, Wind Farm Industry (2009)* 16.
56. Carson, above n. 7, 253.

offshore wind energy could be given priority over other uses,[57] no national SEA in relation to offshore wind energy, or offshore renewable energy more generally, is underway or currently planned. Currently, assessment and approval will take place on an ad hoc basis.

In Australia, the impacts of marine renewable energy projects on the environment and other users will be assessed under general planning and environment laws, as there is no sectoral-specific legislation governing offshore renewable energy projects. The regime for offshore wind energy across Australia involves many different state and federal policies, statutes and approval bodies.

[A] Developments in State/Territory Coastal Waters

[1] Planning Law, Environmental Impact Assessment, and Public Participation

Potential offshore wind energy developments in state coastal waters are governed by state and Territory legislation. Usually, the primary type of authorization required will be a development authorization under state planning law, issued by the relevant statutory authority, government department or Minister for Planning, depending upon the planning procedure. This would be required to construct the offshore wind turbines on the seabed, lay cables to the shore and construct an onshore electrical substation. Currently, there is no preferential scheme for offshore wind energy. Planning/environmental assessment and approval would simply take place within the existing general state planning framework.

For a wind farm proposed to be located in South Australian coastal waters, the *Development Act 1993* (SA) is the primary statute under which environmental assessment of a proposed offshore wind energy facility must be undertaken, and development authorization must be obtained. Development authorization must be obtained for construction of the turbines on the seabed, laying cables, and construction of any substation on land. Currently, the assessment and approval of an offshore wind farm, which is a complex and extremely expensive infrastructure development, is likely to take place under one of the two procedures set out in the Act, these being either the "Crown development and public infrastructure" provisions, or the major developments procedure.[58]

First, the "Crown development and public infrastructure" provisions, set out in section 49 of the *Development Act 1993* (SA), establish a special assessment and approval process for the private sector construction of public infrastructure, which is

57. The European Wind Energy Association has stated that maritime spatial planning is "key to enhancing offshore wind development. It provides stability and clarity for the investors and can bring down the costs of wind energy through an optimum integration of the wind farms into the marine environment": European Wind Energy Association (EWEA), http://www.ewea.org/policy-issues/offshore/ (accessed September 23, 2015).

58. The South Australian Development Act is in the process of being reformed by the *Planning, Development and Infrastructure Act 2016* (SA). Where relevant, proposed changes have been mentioned in footnotes. However, these are not yet in force.

supported by a State agency.[59] This would include the construction of an offshore wind farm to generate electricity and transmission lines to connect the facility directly to the grid. Where such a project is initiated or supported, and endorsed by a State agency, that State agency must lodge the development application with the Development Assessment Commission (DAC).[60]

The DAC assesses the proposal against the relevant development plan, in this case the *Coastal Waters Development Plan*. Section 49(7d) sets out a process for public consultation – which will include industries with competing interests such as commercial fishing – for developments involving construction work totaling more than AUD 4 million, which will apply to large-scale offshore wind farm developments. The DAC must (a) by public advertisement, invite interested persons to make written submissions to it on the proposal within a period of at least fifteen business days; and (b) allow a person who has made a written submission and who, as part of that submission, has indicated an interest in appearing before it, a reasonable opportunity to appear personally or by representative before the DAC to be heard in support of their submission; and (c) give due consideration in its assessment of the application to any submissions.

The DAC must then prepare a report for the Minister for Planning, who is responsible for approving the development. If the DAC believes the development is seriously at variance with the development plan, then specific reference must be made to this fact in the report. If the Minister approves the development, but the DAC believes the development is seriously at variance with the development plan, the Minister must prepare a report and table it before both Houses of Parliament. The report is for noting by Parliament only and does not delay the development. There is no right of appeal for any person against the Minister's decision.[61]

Alternatively, under section 46 of the *Development Act 1993* (SA), the Minister for Planning can declare a proposed development a "major development" where this is appropriate or necessary for proper assessment of the proposed development, and where the proposal is considered to be of major economic, social or environmental importance.[62] Once declared a major development, the application is referred to the

59. Section 49A sets out a procedure for the development of electricity infrastructure not sponsored by a State agency. This approvals procedure largely mirrors the process in s. 49, except the proponent is responsible for submitting the development application. However, as s. 49A does not apply to development for the purposes of the provision of an electricity generating plant with a generating capacity of more than 30 MW (s. 49A(2)), and the average size of an offshore wind farm in Europe in 2015 was over 300 MW (http://www.gwec.net/global-figures/global-offshore/), s. 49A is unlikely to be relevant.
60. The *Planning, Development and Infrastructure Act 2016* (SA) will introduce a new s. 131 to the Development Act, "Development Assessment – Crown development," which basically replicates and replaces the current s. 49 procedure. Section 131 will apply to electricity generation facilities as "essential infrastructure," as defined in s. 3. Applications would be made to the new State Planning Commission. Public rights of participation would apply to facilities with construction costs exceeding AUD 10 million. The *Planning, Development and Infrastructure Act 2016* (SA) repeals s. 49A and does not replace it with an equivalent provision.
61. *Development Act 1993* (SA), s. 49(17).
62. *Development Act 1993* (SA), s. 46. The power of the Minister to call in major developments will continue to operate under the planned amendments to the Development Act.

DAC to determine which type of environmental assessment is required and to issue formal assessment guidelines. The draft environmental impact documentation must include a statement of the extent to which the expected effects of the development are consistent with the provisions of any relevant development plan.[63]

The relevant draft environmental impact documentation is made available for public consultation. The Major Development process is a high-profile, public process during which all key documents are made available to the public. A formal consultation phase is held, during which any member of the public - including industries with competing interests such as commercial fishing - may make a submission on the offshore wind energy development, and a public hearing is held to allow oral submissions to be made. After public consultation, a response is prepared by the proponent.[64] The Minister (with the assistance of the Department of Planning, Transport and Infrastructure) is responsible for assessing the proposal and publishing that assessment in an *Assessment Report*.[65] The Governor of SA (with the approval of Cabinet) will make a decision on the final proposal.[66] Among other things, the Governor must have regard to the provisions of the appropriate Development Plan. Again, no person - including the fishing industry - may appeal against the Governor's decision.[67]

Central to both of these procedures is the *Coastal Waters Development Plan*,[68] against which development applications will be assessed on their merits. This document lists a number of Objectives and Principles of Development Control, designed to provide guidance to decision-makers when assessing whether prospective developments in coastal waters should be approved, and if so, under what conditions. It is in receiving submissions from the public and reports from referrals to other relevant government agencies, and assessing developments against the Objectives and Principles of Development Control in the Development Plan, that the interests of the environment and other users of the marine environment will be taken into account by the decision-maker.

[2] Other Licenses/Approvals

As well as development approval, a range of other approvals will usually be required, including a lease to occupy the sea bed (as Crown land) and an electricity generation

63. *Development Act 1993* (SA), ss. 46B(4)(b), 46D(4)(b).
64. *Development Act 1993* (SA), ss. 46B(5)(b), 46B(8), 46C(5)(b), 46C(8), 46D(5)(b), 46D(7).
65. *Development Act 1993* (SA), ss. 46B(9), 46C(9), s. 46D(8).
66. *Development Act 1993* (SA), s. 48(1).
67. *Development Act 1993* (SA), s. 48(12).
68. South Australian Department of Transport and Urban Planning, *Land Not Within a Council Area (Coastal Waters) Development Plan* (July 4, 2013). This Development Plan applies to the land bounded by the State borders with Western Australia and Victoria, the high water mark along the whole of the South Australian coast and the line 3 nm seaward of the low water mark, and includes both the Spencer Gulf and the Gulf St Vincent, the offshore islands and the land 3 nm seaward of the low water mark around the offshore islands. As noted above, the *Planning, Development and Infrastructure Act 2016* (SA) abolishes Development Plans and introduces new State Planning Policies. These will take some time to draft.

license issued by the state regulator. Various environmental approvals will be required under state and Commonwealth law to protect environment and heritage, usually from the State and/or Commonwealth Minister for the Environment. Legislation exists to protect marine parks and reserves, native vegetation, listed endangered species and migratory species, Aboriginal heritage/artifacts and historic shipwrecks. Consultation will be required under planning law and other legislation in relation to parties with conflicting interests, for example, native title rights and interests; fishing; aquaculture; mining; petroleum (including facilities and pipelines); shipping/navigation and maritime safety; tourism; defense activities and border protection; submarine electrical and telecommunication cables.

For example, for a wind farm proposed to be located in South Australian coastal waters, tenure is secured through a lease or license over the seabed issued under the *Harbors and Navigation Act 1993* (SA).[69] An electricity generation license issued by the Essential Service Commission of SA (ESCOSA) is required to operate in the electricity market.[70] ElectraNet, which is the principal monopoly Transmission Network Service Provider for SA's high-voltage electricity transmission network, and already has a transmission license from ESCOSA, is responsible for connecting wind energy facilities to the grid. Environmental assessment of all aspects of a proposed offshore wind energy facility - including generation and transmission - must be undertaken, and development authorization obtained, from the state Minister for Planning under the *Development Act 1993* (SA).

During the development assessment process under the *Development Act 1993* (SA), referral to the Coast Protection Board under the *Coast Protection Act 1972* (SA) is likely to be required, particularly for onshore substations that impact upon the coastline, and depending on the impact of the proposed wind farm, a license/authorization from the Environment Protection Authority may be required under the *Environment Protection Act 1993* (SA). Approval under the "EPBC Act" may be required to protect nationally and internationally listed threatened species and migratory species found in SA's coastal waters and on the coastline.[71] A permit may be required under the *National Parks and Wildlife Act 1972* (SA), which extends protection of state listed areas and threatened species under this Act to coastal waters. The *Fisheries Management Act 2007* (SA) protects aquatic reserves established under the Act. Approval to clear native vegetation under the sea or on the coast may be required under the *Native Vegetation Act 1991* (SA).

Laws for the protection of indigenous heritage under the *Aboriginal Heritage Act 1988* (SA) and State heritage (buildings, jetties) also apply to coastal areas and may be

69. In Australia, Commonwealth legislation gives jurisdiction over the seabed under state coastal waters to the Crown in right of the State: *Coastal Waters (State Title) Act 1980* (Cth). In SA, all subjacent land is vested in the Minister for Transport under the *Harbors and Navigation Act 1993*, and under the Minister's care, control and management. Prior to the commencement of the *Harbors and Navigation Act 1993* (SA), the seabed below the high water mark to the State limit was Crown land under the *Crown Lands Act 1929* (SA) (repealed by the *Crown Land Management Act 2009* (SA)), and under the administration of the Minister for Environment and Conservation.

70. *Electricity Act 1996* (SA).

71. For a discussion of this Act, *see* section §7.04[B], below.

relevant depending upon the location of substations onshore. Shipwrecks are protected under the *Historic Shipwrecks Act 1981* (SA). Consultation and negotiation with Indigenous peoples who hold native title rights and interests over offshore areas or in regards to the coast is required under the *Native Title Act 1993* (Cth).

Finally, the *Fisheries Management Act 2007* (SA) protects fisheries established under the Act, while aquaculture is protected under the *Aquaculture Act 2001* (SA). The rights of petroleum and offshore minerals companies are protected through tenements granted under the *Petroleum (Submerged Lands) Act* 1982 (SA) and *Offshore Minerals Act 2000* (SA). Consultation may be required under the *Defence Act 1903* (Cth). Shipping and navigational interests, and submarine existing electrical and telecommunication cables, are also protected under international and national law.

Although the various state planning procedures may be lengthy, complex, and are not necessarily adapted to all the particular issues facing offshore wind energy, existing state legislation is sufficient to deal with the assessment and approval of offshore wind energy where a wind farm is proposed for state coastal waters.

[B] Developments in Commonwealth Coastal Waters

In contrast, where a wind farm is proposed in Commonwealth waters, the Commonwealth regime is inadequate. There is no specific legislation by which tenure for offshore wind energy developments in Commonwealth waters may be obtained, for example, through a wind energy license or lease, or which establishes a specific procedure for the assessment and approval of offshore wind energy facilities in Commonwealth waters.

Prior to 2014, it appeared the installation and operation of offshore wind turbines would fall within the scope of the *Sea Installations Act 1987* (Cth), which made it an offence to install, use or carry out certain work while on a sea installation, in waters from 3 nm to 200 nm seaward, without a permit.[72] The Act placed various restrictions on the operation of sea installations, for example, that activities would not interfere with navigation, fishing or the conservation of the resources of the sea or seabed.[73] However, the offence and permitting provisions were repealed in 2014, as part of a general omnibus repeal of legislation. The Act still empowers the Minister to establish safety zones around sea installations and direct that sea installations be removed.[74]

The main procedure relevant to assessing offshore wind farms in Commonwealth waters is contained within the EPBC Act, which establishes a system for Commonwealth environmental assessment and approval of "controlled actions." The Act prohibits a person from undertaking a controlled action unless certain processes have been followed and/or certain approvals have been obtained. The most common

72. *Sea Installations Act 1987* (Cth), ss. 14-16 (repealed). *See* A Bradbrook and A Wawryk, "The Legal Regime Governing the Exploitation of Offshore Wind Energy in Australia" (2001) 18 *Environmental Planning and Law Journal* 30, 40-42.
73. *Sea Installations Act 1987* (Cth), s. 23 (repealed).
74. *Sea Installations Act 1987* (Cth), ss. 51, 57 and 55.

controlling provisions (or "triggers") are the nine "matters of environment significance" set out in the Act. Of these, section 23 is the most relevant for proposed offshore wind energy facilities.[75] Under this section, a person must not take in a Commonwealth marine area an action that has, will have or is likely to have a significant impact on the environment. Also, a person must not take outside a Commonwealth marine area but in the Australian jurisdiction an action that has, will have, or is likely to have, a significant impact on the environment in a Commonwealth marine area.[76]

It is the responsibility of the proponent to seek a determination from the Commonwealth Environment Minister as to whether the development is a controlled action. The EPBC Act does not require referral, assessment and approval of every proposed development, only those actions that have, will have or are likely to have a significant impact on a matter of environment significance. However, given the potential significant environmental impacts of commercial offshore wind farms, it is likely referral will be required.[77] If the Minister decides the action is a controlled action, the proponent must conduct an environmental assessment of the development under one of the assessment procedures set out in the Act.

The Commonwealth Environment Minister is responsible for approving developments, although his or her powers are delegated to the Commonwealth Environment Department. When making assessment and approval decisions under the EPBC Act, the Minister must "have regard" to the provisions of any of the five marine bioregional plans which may be relevant. These plans, which provide guidance to decision-makers about conservation priorities in each of the five marine planning areas, or bioregions, established under the EPBC Act,[78] may thus significantly influence future offshore wind energy development activity in the Commonwealth marine area.

An offshore wind energy proponent may also need a permit from the Director of National Parks to undertake certain activities in one of the marine protected areas – called Commonwealth marine reserves – established in the Commonwealth marine

75. Other controlling provisions of potential relevance to proposed offshore renewable energy facilities are: s. 18 (prohibiting an action that has, will have, or is likely to have, a significant impact on a listed threatened species) and s. 20 (prohibiting an action that has, will have, or is likely to have a significant impact on a listed migratory species). Furthermore, Ch 5 Pt 13 prescribes criminal penalties for the following offences undertaken without a permit: the killing, injuring or taking of listed endangered species (ss 196–196E); the killing, injuring or taking of listed migratory species (ss 210–212); and the killing, injuring or taking of whales and other cetaceans (ss 224–236).

76. The Commonwealth marine area is defined in s. 24. This includes any waters of the sea inside the seaward boundary of the exclusive economic zone, and any waters over the continental shelf, except State or Territory waters. The Commonwealth marine area thus covers waters seaward from 3 nm from the coast to the boundary of the exclusive economic zone (200 nm), and includes the seabed under, and airspace over, these waters.

77. The *Wind Farm Industry Policy Statement* (2009) identifies the possible significant environmental impacts on the Commonwealth marine area: *see* above n. and accompanying text.

78. Excluding the Great Barrier Reef Marine Park, which is zoned and managed through its own legislation.

area under the EPBC Act.[79] Historic shipwrecks in Commonwealth waters are protected under the *Historic Shipwrecks Act 1976* (Cth).

Any proposal to establish a wind energy facility in Commonwealth waters must consider the rights of other users in Commonwealth waters. The *Offshore Petroleum and Greenhouse Gas Storage Act 2006* (Cth) and *Offshore Minerals Act 1994* (Cth) regulate the exploration and production of petroleum and minerals, and the rights and interests of resources developers are protected by various tenements granted under these Acts. The *Fisheries Management Act 1991* (Cth) regulates fisheries in the Australian Fishing Zone (waters seaward from 3 nm), including the grant of statutory fishing rights. Finally, because the Commonwealth has exclusive legislative power over some matters, such as quarantine, defense, and immigration, Commonwealth jurisdiction over Commonwealth offshore areas raises issues not relevant to the States. Defense facilities and activities may constitute a conflicting use of ocean space for offshore wind energy facilities.

The lack of law and policy guiding offshore wind energy developments in Commonwealth waters is a continuing deficiency in the legal regime.[80] There is no clear existing statutory process addressing the acquisition of tenure and establishing an assessment and approval process. Furthermore, there is a singular lack of clarity and certainty on how an offshore wind energy project application for a wind farm that spans Commonwealth and State waters will be administered, for example, where turbines are sited in Commonwealth waters, with undersea cables running from Commonwealth through State waters to the coast, and a substation constructed onshore; and/or where turbines spanning State and Commonwealth waters. The turbines placed in state coastal waters will require rigorous assessment and approval and potentially require referral to a number of state agencies; while the turbines located in Commonwealth waters will not. It is not clear which agency (if any) will have overall responsibility for tenure allocation and assessment of the project.

[C] Compensation

The development of offshore wind energy may have negative impacts on other users of the sea, including the fishing industry. Offshore development may affect fish resources,

79. *Environment Protection and Biodiversity Conservation Act 1999* (Cth) s. 344. The EPBC Act and the EPBC Regulations 2000 specify a range of activities that may be carried out in Commonwealth reserves. Some activities - including excavations, erecting a building or other structure, carrying out works, taking an action for commercial purposes and taking a native species - must be carried out in accordance with a management plan for the reserve: *Environment Protection and Biodiversity Conservation Act 1999* (Cth), ss. 354, 354A, 355 and 355A. Other activities, such as the use of vessels, are either prohibited outright by the Regulations, or prohibited unless a permit is obtained or they are approved by a management plan: *Environment Protection and Biodiversity Conservation Regulations 2000*, Pt 12. Also, the Director of National Parks may control activities in Commonwealth reserves through determinations, prohibitions, or restrictions made under the Regulations: *Environment Protection and Biodiversity Conservation Regulations 2000*, Regs 12.23 and 12.23A.
80. A Wawryk, "Legislating for Offshore Wind Energy in South Australia" (2011) 28 *Environmental and Planning Law Journal* 265.

causing loss of catch and fishing income, while the declaration of safety exclusion zones means fishing may be excluded in traditional fishing grounds. The issue of compensation is thus one that has arisen in other jurisdictions.

Should an offshore wind energy facility be granted development authorization in Australia, there is no right to compensation for loss suffered by other industries under planning law. For example, there are no provisions within the South Australian *Development Act* by which compensation would be payable by offshore wind energy developers to industries such as commercial fishing. This is because development law assumes that planning permission is given only after the costs and benefits of the development have been assessed by the decision-maker, and that permission is only given where the use of the land/waters is appropriate. Planning law seeks to assess, manage and mitigate impacts, and once authorization is given, individuals who suffer loss are not entitled to compensation under planning law.

There are very complex principles of law regarding compensation for the acquisition of property rights, and their application to commercial fishing is not clear. There is a presumption at common law that compensation is payable when the government takes away property.[81] However, as a presumption, this can be rebutted by clear legislative intent to the contrary. Thus, even if fishing entitlements are property – which is unlikely since the High Court case *Harper v. Minister for Sea Fisheries*[82] – if there is a clear legislative intent that parliament intended to take the right away without paying compensation, then there will be no legal avenue by which a fishing rights-holder can seek compensation at common law.[83]

As a matter of constitutional law, the state governments are not required to pay compensation for the deprivation or acquisition of property rights. Thus, if the approval of an offshore wind farm in state coastal waters were to interfere with fishing rights, the South Australian government has no obligation to pay compensation.

In contrast, section 51(xxxi) of the Commonwealth Constitution provides that the Commonwealth Parliament has the power to make laws for the peace, order and good government of the Commonwealth with respect to the "acquisition of property on just terms from any State or person for any purpose in respect of which the Parliament has the power to make laws." This is seen as a constitutional guarantee for the payment of just compensation for the compulsory acquisition of property by the Australian Government. However, jurisprudence to date suggests that is seems unlikely that the approval of an offshore wind farm (presumably pursuant to enabling Commonwealth legislation) will be characterized as an "acquisition of property." First, it is questionable whether statutory fishing rights created under the *Fisheries Management Act 1991* (Cth) would be "property" for the purposes of section 51(xxxi), and second, it would be very difficult to argue there has been an "acquisition" of a property right.[84]

---

81. *Attorney-General v. De Keyser's Royal Hotel* [1920] AC 508 at 542 (Lord Atkinson), 579 (Lord Parmoor).
82. (1989) 168 CLR 314.
83. W Gullet, *Fisheries Law in Australia* (*LexisNexis Butterworths*, 2008), 71.
84. *Ibid.*

Although it would be possible for the Commonwealth and State governments to enact legislative provisions requiring the payment of compensation to the commercial fishing industry, current fisheries legislation does not give fishing rights holders any right to compensation for loss caused by ocean developments.[85] Precedent to enact such provisions could be found in mining law, where courts can require mining companies to pay compensation to landowners for loss or damage, including "economic loss, hardship or inconvenience" caused by mining activities on private land.[86] However, it would not be consistent with general planning law. It is probably more likely to expect a voluntary approach to the payment of compensation by offshore wind energy developers, as has occurred in the United Kingdom.

§7.05 GRID ISSUES

Historically, each Australian state and Territory had its own State/Territory electricity supply industry, with generation, transmission and distribution under the control of a state-owned monopoly. From the 1990s, these state-owned monopolies were broken into separate arms, so that now generation and retail activities are competitive, while transmission and distribution are regulated monopolies. In 1996, a legislative framework for the creation of a NEM was introduced, and one NEM exists between the States of SA, Queensland, New South Wales, Victoria and Tasmania. Because of the distances involved, the state of Western Australia and the Northern Territory are not part of the NEM.

The development of the NEM was built on a framework geared to the existence of large conventional power stations - typically coal-fired power stations - constructed close to coastal population centers, with access to transmission lines constructed by the state when the industries were government-owned, vertically integrated monopolies. The NEM rules, including those for access to the grid, have been largely built around liberalizing and introducing competition to market(s) once dominated by the state-owned electricity companies, to ensure the efficiency of electricity supply.

Although a detailed discussion of the NEM rules is beyond the scope of this chapter, it can be said that renewable energy is disadvantaged by the existing grid layout, and network access and investment rules, which favor preexisting conventional large coal-fired power stations. Australia's grid is one of the longest in the world, running coast-to-coast for some 5,000 km, with limited interconnections between the states. The need to construct new electricity transmission infrastructure for offshore wind, and limited interconnections, which are necessary to export wind energy to other regions when supply is high, may constrain the development of offshore wind energy. In states such as SA, where wind energy has already achieved significant penetration, "upgrades and extensions to the current grid may be needed to accommodate significant further wind energy development."[87] However, under the NEM rules, network access costs such as investment in new infrastructure will usually be borne by the

85. *Fisheries Administration Act 1991* (Cth); *Fisheries Management Act 2007* (SA).
86. *See*, for example, the *Mining Act 1971* (SA) subs 9AA(9), s. 61.
87. Carson, above n. 7, p. 240.

offshore wind energy proponent, making this a cost barrier to entry.[88] There are no preferential rules for offshore wind energy.

§7.06 CHALLENGES AND SOLUTIONS

The Australian system of law governing developments in offshore waters is complex, lacks clarity and is not nationally consistent. Because of this, it has been argued the legal regime acts as a barrier to the development not only of offshore wind energy but also for other forms of marine energy such as tidal and wave power.[89]

One of the gaps in the legal regime that must be addressed is the issue of tenure for offshore wind energy projects in Commonwealth waters. In order to provide clarity, certainty and transparency for the industry and community, the introduction of a sectoral-specific licensing regime for offshore wind energy should be considered. In particular, the development of a joint Commonwealth-State regime governing the assessment and approval of offshore renewable energy developments could provide security of tenure as well as allowing environmental impacts in both Commonwealth and State offshore area to be assessed and considered holistically. These are suggestions have been addressed elsewhere in more detail, and remain relevant today.[90]

Second, the federal and governments should consider conducting SEA to assess the environment of marine areas, identify competing interests in those areas, and provide information to help determine areas where offshore wind energy could be given priority over other uses.[91] SEA requires a thorough, transparent and time-consuming process, requiring the commitment of significant investment and resources,

88. For a discussion of the disadvantages faced by renewable energy generators in the National Electricity Market in Australia in relation to matters such as network access, *see* L Godden and A Kallies, "Electricity Market Development: New Challenges for Australia" in M Roggenkamp, L Barrera-Hernández, D Zillman and I del Guayo (eds) *Energy Networks and the Law* (Oxford University Press, 2012) 292; L Byrnes, C Brown, J Foster and L Wagner, "Australian Renewable Energy Policy: Barriers and Challenges" (2013) 60 *Renewable Energy* 711.
89. D Leary and M Esteban, "Recent Developments in Offshore Renewable Energy in the Asia-Pacific Region" (2011) 42 *Ocean Development and International Law* 94; D Leary and M Esteban, "Climate Change and Renewable Energy from the Ocean and Tides: Calming the Sea of Regulatory Uncertainty" (2009) 24 *The International Journal of Marine and Coastal Law* 617; D Leary, "Planning Law Challenges and Options for Renewable Energy", Paper presentation, *Workshop: Challenges and Opportunities for Renewable Energy in South Australia,* Renewables SA in conjunction with the University of New South Wales.

    Centre for Energy and Environmental Markets (May 10, 2010); G Wright, "Marine Energy in Australia and New Zealand: Regulatory Barriers and Policy Measures", Paper presentation, All-Energy Australia Conference, Melbourne 2011; G Wright, "Marine Renewable Energy: Legal and Policy Challenges to Integrating an Emerging Renewable Energy Source", Paper presentation, IKEM International Summer Academy on Energy and the Environment, Berlin 2012.
90. Above nn [80][89].
91. The European Wind Energy Association has stated that maritime spatial planning is "key to enhancing offshore wind development. It provides stability and clarity for the investors and can bring down the costs of wind energy through an optimum integration of the wind farms into the marine environment": European Wind Energy Association (EWEA), http://www.ewea.org/policy-issues/offshore/, accessed September 23, 2015.

and community consultation.[92] However, it has a number of advantages, including more strategic development along the coast.[93] SEA can provide certainty that projects will be located in predefined areas with minimal environmental impacts and multiple-use conflicts, providing assurance to communities and interest groups that "they will not need to repeatedly engage in time-consuming and costly disputes with what may prove to be a long series of project proposals."[94] Also, the costs for project proponents are reduced, as the areas for investigating prospective wind energy sites are already narrowed.

Third, the various Australian governments should consider how rights in offshore wind energy should best be allocated. Governments may choose to retain the current ad hoc, individual application process, whereby individual companies identify potential sites for projects and approach the various approvals authorities to obtain the necessary permissions to construct and operate the facility, if they are of the opinion development is commercially viable. An individual procedure offers flexibility to project proponents. Alternatively, governments may introduce a competitive process for allocating rights, such as a reverse-auction or tender system, where the government releases predefined areas and/or specified projects, and businesses compete to be awarded the rights to develop the offshore wind energy facility. The winning company may be granted not only the right to develop a wind energy facility in a certain predefined area or to develop a specified project but also an incentive such as a power purchase agreement with electricity utilities and/or a feed-in tariff.

While there are various costs and benefits of each process, competitive processes are increasingly being used for renewable energy developments, both onshore and offshore, because of their transparency and to ensure facilities are developed at a competitive cost.[95] By early 2015, sixty countries had adopted renewable energy auctions.[96] Renewable energy auctions for various renewable energy projects were used in twelve of the twenty Latin American countries.[97] European countries such as the UK, Denmark, the Netherlands and Germany have introduced, or are in the process

92. Department of Sustainability and Environment (State of Victoria), *Marine Energy in Victoria: Discussion Paper* (2010) 45.
93. *Ibid.*
94. *Ibid.*
95. *Ibid*, 33. For a discussion of the matters to be addressed under a competitive procedure and the advantages and disadvantages of competitive procedures, *see also* IRENA & CEM, *Renewable Energy Auctions: a Guide to Design 1* (2015), http://www.cleanenergyministerial.org/Portals/2/pdfs/IRENA_RE_Auctions_Guide_2015_1_summary.pdf, accessed July 26, 2016; IRENA & CEM, *Renewable Energy Auctions: A Guide to Design 2* (2015), http://www.irena.org/DocumentDownloads/Publications/IRENA_RE_Auctions_Guide_2015_2_policies.pdf, accessed July 26, 2016.
96. IRENA & CEM, *Renewable Energy Auctions: A Guide to Design 1* (2015), *ibid.*, 13.
97. IRENA, *Renewable Energy Policies in Latin America 2015: An Overview of Policies* (June 2015) p. 11, Table 1. For case studies of renewable energy auctions in Brazil, China, Morocco, Peru and South Africa, *see* IRENA, *Renewable Energy Auctions in Developing Countries* (2013) https://www.irena.org/DocumentDownloads/Publications/IRENA_Renewable_energy_auctions_in_developing_countries.pdf.

of introducing, competitive procedures for offshore wind energy.[98] The Australian Capital Territory has used reverse auctions for large-scale onshore wind energy generation projects, awarding a feed-in tariff entitlement to the successful proponent.[99] However, auctions are appropriate when there are multiple interested bidders, a situation which seems unlikely to exist in Australia in relation to offshore wind energy in the near future.

Finally, renewable energy developers and investors require certainty in the law, not only in relation to planning law, but also in relation to policies and measures supporting the development of renewable energy. In Australia, there has been considerable uncertainty in relation to energy policy over recent years. The rules of the RET have been reviewed every two years and changed numerous times, which combined with a lack of bipartisan support for the legislation, and the introduction and abolition of the CPM within two years, has created uncertainty and undermined investment in renewable energy projects.[100] In recognition of this, biennial reviews of the RET were abolished in 2015. The ongoing differences between the two major political parties over pricing GHG emissions, the RET, the Safeguard Mechanism and emissions trading, will not assist in providing a stable environment for large-scale investments in offshore wind energy. A bipartisan approach to energy policy is required, although this seems unlikely to be achieved in the near future.

§7.07 CONCLUSION

The Australian government has recently expressed some policy support for offshore wind energy, through its inclusion as a technology to be supported through the new CEIF. Existing mechanisms such as the RET, and the new Safeguard Mechanism, may help to make the costs of offshore wind more competitive with fossil fuel-sourced electricity, although electricity network access costs may act as a barrier to offshore wind development, as the proponent usually bears the cost of access to the grid. Continuing uncertainty and a lack of a bipartisan approach to energy policy may undermine certainty and investment in large-scale capital enterprises such as offshore wind energy.

Assessment and approval processes for offshore wind energy projects in state coastal waters are lengthy and complex, as the competing interests of user/environment and multiple users must be comprehensively assessed and managed, while the regulatory regime for developments in Commonwealth waters is deficient, lacking a clear and transparent licensing regime. The Australian state and

---

98. European Wind Energy Association, *Design Options for Wind Energy Tenders* (December 2015) http://www.ewea.org/fileadmin/files/library/publications/position-papers/EWEA-Design-opt
ions-for-wind-energy-tenders.pdf, accessed July 26, 2016.
99. *Electricity Feed-in (Large-scale Renewable Energy Generation) Act 2011* (ACT).
100. Australian Government, Climate Change Authority, *Renewable Energy Target Review: Report* (December 2015), 50 http://www.climatechangeauthority.gov.au/reviews/2014-renewable-energy-target-review, accessed July 27, 2016.

federal governments should consider introducing sectoral-specific legislation to establish a clearer and more certain licensing process.

A government committed to developing a large-scale offshore wind energy industry may consider bearing some of the costs and reducing the uncertainties faced by developers, by undertaking SEA. As potential conflicts are addressed as part of the SEA process, the approval processes for individual applications in priority areas may be simpler and quicker. After conducting SEA, governments could consider the costs and benefits of maintaining an ad hoc individual assessment and approval regime, or implementing a competitive procedure for the allocation of rights. The experience of other jurisdictions in auctions and tenders could offer useful guidance in this regard.

CHAPTER 8

Legal Framework to Develop Offshore Wind Power in China

Haifeng Deng

§8.01 INTRODUCTION

By 2012, the first commitment period of the Kyoto Protocol has expired. According to the Paris agreement reached at the 21st Conference on Climate Development, countries have submitted their own emission reduction commitments. The developed countries proposed their emission reduction targets using 1990 or 2005 as the baseline. Like USA greenhouse gas emissions reduced to 25% by 2025, which represents a cut on 2005 levels. And the developing countries put forward their own mitigation goals, which focus on the decline in emissions. China has pledged to slow down the speed of its growing carbon emissions by 40% to 45% from 2005 levels by 2020.[1] In the "Twelve Five-Year Plan" for renewable energy development, the government has made the target to achieve the total generating capacity of more than 20% by 2015.[2] During the "Thirteen Five-Year Plan" period, the total investment in wind power construction will reach RMB 700 billion, and by the end of 2020, wind power generation will be ensured to achieve 420 billion KW hours, accounting for about 6% of the country's total generating capacity.[3] Achieving this goal is a challenge for China. China's energy structure reform is imperative.

Compared with other new energy sources, wind power has a short construction period, put into operation quickly. Until 2005, the nation started the offshore wind

---

1. *Paris Agreement to the United Nations Framework Convention on Climate Change* (2015), opened for signature April 22, 2016, FCCC/CP/2015/L.9 ("*Paris Agreement*").
2. *The 12th Five-Year Plan for Economic and Social Development (2011-2015)*, available at http://www.gov.cn/2011lh/content_1825838_2.htm.
3. *The 13th Five-Year wind power development planning*, available at http://www.gov.cn/xinwen/2016-11/30/content_5140637.htm.

energy included in the development directory. In 2007, China's first offshore wind turbine in the Bohai Sea was put into operation, and the power was put into the Bohai Oilfield independent power grid. In June 2010, China built the first large offshore wind farm – Shanghai Donghai Bridge 10.2 * 10^4 KW offshore wind power demonstration project, which became the world's first offshore wind farm in addition to Europe. In September 2010, China's first offshore wind power concession project tender by the National Energy Administration organized to determine the total size of 100 * 10^4 KW four projects.[4] By the end of 2014, China has completed a total of 114,609 MW offshore wind power projects.[5] In addition to the already completed offshore wind power projects, there are a number of projects for which the National Energy Administration has agreed to carry out preliminary work of offshore wind power from the beginning to the scale of the construction of a critical stage. By the end of 2015, the national offshore wind power grid reached a capacity of 750,000 KW; compared with the development of onshore wind power, the development of offshore wind power is still the problem of wind power industry itself, and more of the technical level of the industry itself, including whether the unit technology, construction technology, transmission technology, operation and maintenance technology to meet the needs of offshore wind power development.[6]

So why the construction of the wind power moves from onshore to offshore? First is about the grid conditions. China's land wind energy resources are mainly distributed in the western region, but this region is the end of the powder grid, and the powder grid construction is poor. In the meanwhile, the power produced in western region will be transported to the southeast coastal areas, which have the most power and the best power grid construction. In order to rational and effective use of wind energy resources to promote the healthy development of wind power in China to ensure grid security and reliable operation, it is necessary to move from the west region to the southeast coastal areas. Second is about the impact on the environment. The onshore wind farm not only sited on the inaccessible wastelands in Xinjiang or Inner Mongolia, but also on the populated areas like Jiangsu and Shanghai. The onshore wind farm will take up the land area and make huge noise.

In 2005, "Eleventh Five-Year Plan" for renewable energy development was put forward to exploring offshore wind power development. The same year, the national development and reform commission listed the offshore wind turbine technology research and development projects in the development guidance catalogue of the renewable energy industry. In 2009, National Energy Administration published "Work Outline for Offshore Wind Farm Project Planning," starting the official work of the offshore wind farm. In 2010, the National Energy Administration and the State Oceanic

4. *State Oceanic Administration. Bulletin of Marine Economy in China in 2012 (2013).*
5. T XU. *China wind power installed capacity of 2014.* China Agricultural Machinery Industry Association, Wind Energy Equipment Branch wind energy industry. (徐涛.(2015).2014 中国风电装机容量统计. 中国农机工业协会风能设备分会风能产业.).
6. Y XIAO. (2016). *From the outbreak of growth to the gradual wind power will bid farewell to the "vase". Energy Research and Utilization* (6).(&#x筱阳. (2016). 从爆发增长到循序渐进海上风电将告别" 花瓶". 能源研究与利用(6)).

Administration jointly issued the "Interim Measures for Management of the Development and Construction of Off-Shore Wind Power."[7]

The primary purpose of this chapter is to give a preview about the construction of China's offshore wind farm in law and policy level in order to solve the challenges in institutional design and regulations in practice. This chapter is composed of five sections including a short Introduction of China's offshore wind farm policy and Conclusion. Section §8.02 discusses the institutional design for offshore wind farm by analyzing the roles of the main authority, supplementary authorities and the law. Section §8.03 will analyzes the legal design for offshore wind, including incentives and regulations. Section §8.04 discusses the challenges and solutions for institutional design, incentives and the regulations.

§8.02 INSTITUTIONAL DESIGN FOR OFFSHORE WIND

[A] The "Main" Authority During the Development of Offshore Wind Power

"Interim Measures for Management of the Development and Construction of Off-Shore Wind Power", jointly issued by the National Energy Administration and the state oceanic administration, is the most important policy in offshore wind farm construction recently. The Article 4 of the paper has stipulated that the national energy authority is responsible for the nationwide management of offshore wind power development and construction. Energy authorities of coastal provinces under the guidance of the national energy authority are responsible for local management in the development and construction of offshore wind power.[8] To further improve the offshore wind power management system, regulate the development and construction of offshore wind power order to promote the sustained and healthy development of offshore wind power industry, the State Oceanic Administration and the National Energy Administration, jointly issued the "Administrative Measures for the Development and Construction of Offshore Wind Power"[9] this year. This document guides the current provisions of China's offshore wind power: at the central government level, the management duties of offshore wind power are shared by the National Energy Administration, as a part of the National Development and Reform Commission and the State Oceanic Administration; at the local government level, the local Energy

7. J He, J Zhao, & J Yang. (2011). *Countermeasures and suggestions for development of offshore wind powder industry in China. Yangtze River,* 42(1), 37–39. (何杰, 赵鑫, & 杨家胜. (2011). 我国海上风电产业发展的对策与建议. 人民长江, 42(1), 37–39).
8. *Notice of the National Energy Administration and the State Oceanic Administration on issuance of the Interim Measures for Management of the Development and Construction of Off-Shore Wind Power* (January 22, 2010), available at http://www.lawinfochina.com/display.aspx?lib=law&id=9120&CGid= (accessed by January 31, 2017).
9. *Notice of the National Energy Administration and the State Oceanic Administration on issuance of the Administrative Measures for the Development and Construction of Offshore Wind Power* (December 29, 2016), available at http://www.soa.gov.cn/zwgk/gfxwj/hygl/201701/t20170110_54446.html (accessed by January 19, 2017).

Authorities and the local Ocean Authorities are responsible for the management of offshore wind power projects.

[1] Central Government Level

It is stipulated in the "Administrative Measures for the Development and Construction of Offshore Wind Power," that the relevant administrative departments at all levels have their own function and responsibility, but the National Energy Administration and the State Oceanic Administration are playing the most important roles in the government, which is basically written in the file.

The National Energy Administration is responsible for the development and management of the national offshore wind power development. The local energy authorities shall be responsible for the development and management of offshore wind power in the region under the guidance of the State Energy Bureau and the renewable energy technology support units do offshore wind power technical services.[10] Marine administrative departments are responsible for offshore wind power development and construction of sea area island use and environmental protection management and supervision.[11]

Specifically, the National Energy Administration unified organization of the national offshore wind power development planning and management with the State Oceanic Administration to jointly test the local offshore wind power development plan; this timely organization of the relevant technical units has helped assess the development of local offshore wind power.[12]

There are some departments in the central government level playing important roles in the offshore wind farm legal regime.

[2] The Role of the Local Government

The local Energy Authorities shall organize relevant units to prepare the offshore wind power development plans within the waters of the local in accordance with the standards and implement the scheme of grid access and market consumption.[13] The local Energy Authorities and the local Ocean Authorities shall carry out the rolling adjustment work of the offshore wind power development plan according to the relevant policies of the national renewable energy development and the development of the offshore wind power industry. The specific procedures shall be carried out according to the planning requirements.[14] The local Ocean Authorities shall put forward preliminary opinions on the preliminary examination of the sea island and the environmental impact assessment (EIA) of the offshore wind power development plan in the region in accordance with the planning of marine main functional areas, marine

10. *Administrative Measures for the Development and Construction of Offshore Wind Power* Art. 4.
11. *Ibid.* Art. 5.
12. *Ibid.* Art. 8.
13. *Ibid.* Art. 9.
14. *Ibid.* Art. 12.

functional zoning, island protection planning and marine economic development plans.[15]

China has not yet developed a national wind power regional planning; domestic enterprises rely mainly on local resources, so many coastal provinces have developed their own offshore wind power development planning, such as the Jiangsu Province's Report, "Ten million kilowatts of wind power base planning report" on the details to develop a power generation plan in Jiangsu Province.[16] Compared with the fiery desire of the wind power developers, some local governments have a relatively negative attitude. Since the wind power is renewable energy, it enjoys the national policy concessions, value-added tax reform, and the wind farm equipment inputs are tax deductible; so during the first seven years of the wind farm construction and operation, the local government almost cannot get tax, which means that local interests in the offshore wind power is not too much reflected.

[3] Other Players Vital for the Development of Offshore Wind

The national energy wind power technology R & D Center approved by the National Energy Administration was formally established. Relying on the largest wind power development enterprise in China – Longyuan Electric Power Group Co., Ltd., the R & D center is based on various types of wind farms all over the country, focusing on the whole industrial chain of wind power and the national energy development strategy,[17] wind farm development and operation, offshore wind farm operation and maintenance, and other key technologies to comprehensively enhance the design, construction and operation and management level of wind farms in China, which are suitable for China's environmental characteristics and topographic conditions. The R & D Center has carried out a series of major R & D projects around the operational support of wind farms.

[B] The Role of "Law" under the Institution Decision Making and Related Policies

When we talk about offshore wind power, there are two important regulations in the area of offshore wind power, "Work Outline for Offshore Wind Farm Project Planning" and "Administrative Measures for the Development and Construction of Offshore Wind Power." And there are other local regulations which could be used. So basically regarding the offshore wind power, most regulations belong to the administrative regulations and local regulations, which limit the development of offshore wind power

15. *Ibid.* Art. 10.
16. J He, X Zhao & J Yang. (2011). *Countermeasures and suggestions for development of offshore wind powder industry in China. Yangtze River, 42(1), 37–39.* (何杰, 赵鑫, & 杨家胜. (2011). 我国海上风电产业发展的对策与建议. 人民长江, 42(1), 37–39).
17. The National Energy Administration: National Energy Research and Development Center (Laboratory) parade of work, available at http://www.nea.gov.cn/2012-04/28/c_131558419.htm (accessed by January 28, 2017).

in China. There are also laws which are not that related but still useful: Marine Environment Protection Law of the PRC, Electric Power Law of the PRC, Law of the PRC on Administrative Permission, Law of the People's Republic of China on the Administration of the Use of Sea Areas, Energy Conservation Law of the PRC, Renewable Energy Law of the PRC, Law on Environmental Impact Assessment of the PRC; and there are administrative regulations: Regulation on Environmental Impact Assessment of Planning, Provisions Governing the Laying of Submarine Cables and Pipelines, Administrative Regulation on the Prevention and Treatment of the Pollution and Damage to the Marine Environment by Marine Engineering Construction Projects.

Law is a system based on any of the policy. Renewable Energy Law is the main law for the offshore wind farm, supplying the direction guidance for the policy making. Offshore wind is a kind of new energy program, so it is always policy guidance, because of the scientific uncertainty. The law has the basic guidance for the policy making, and on the other hand, most of the policies are not specific as law. So, separate offshore wind power laws also have a lot of room for further development.

§8.03 LEGAL DESIGN FOR OFFSHORE WIND

[A] Incentives

[1] Main Finance Scheme

There are incentives for renewable energy in China, like renewable energy total target system, the compulsory system of renewable energy power generation, classification tariff system, tax incentives, fiscal investments and subsidies. But there is only feed-in tariff policy for the offshore wind farm. On June 19, 2014, the National Development and Reform Commission issued "on the offshore wind power tariff policy" notice, which is an offshore wind power tariff policy to encourage the development of high quality offshore wind power resources. Through the concession tender to determine the owners of offshore wind power projects, the FIT must be performed in accordance with the bid price, but not higher than the same level of government pricing. Although the notice clearly applies to the time limit as 2017, it does not include the offshore wind power project put into operation in 2017.[18]

One of the biggest difficulties in the commercialization of new energy is the price which is higher than the traditional energy. So, it is inevitable that the state has to take the form of policy subsidies, as at present our country make a universal payment for the development of new energy; the existing electricity price includes renewable energy of the additional price. Offshore wind power development is much more difficult than onshore wind power; its power generation technology is behind land-based wind power for about ten years, the cost should be two to three times higher. China's first offshore wind power demonstration project – Shanghai Donghai Bridge offshore wind

18. *Notice of the National Development and Reform Commission on issuance of the on offshore wind power price policy* (June 5, 2014), available at http://www.sdpc.gov.cn/gzdt/201406/t20140619_615709.html.

farm project, after-tax electricity price of CNY 0.978 per KW-hour.[19] Since 2017, the offshore wind power projects have been put into operation; offshore wind power technology will be based on progress and project construction costs changes, combined with the situation of concession bidding to study the development of electricity price policy. This encourages the adoption of concession bidding and other market competition to determine the development of offshore wind power projects and the owners of electricity prices to promote technological progress. The concession bidding shall determine the owner's offshore wind power project. The on-grid electricity price shall be executed according to the bid price, but shall not be higher than the price level of similar projects.

[2] Supplementary Finance Scheme

According to the "Notice on declaration of foreign government loan alternative project in 2015," some foreign government loans can be used for China's clean energy and renewable energy project needs.[20] The People's Bank of China has launched green financial bonds in the inter-bank bond market to support the green industry and according to the agreed debt service of securities. There is no special loan policy for the offshore wind farm. According to the "National carbon emissions trading market notification of commencement of priorities", the power has been put into carbon emissions trading market. So the offshore wind power could be traded in the carbon emissions trading market.

The "Renewable Energy Law" and the "the Interim Management Measures for the Renewable Energy Special Fund" do not specifically protect the local offshore wind power industry measures, but Article 24 of the Renewable Energy Law promotes the development and utilization of renewable energy equipment of the localization of production content, in a certain level also through the funding to subsidize the local industry.[21] According to Article 26 of the Renewable Energy Law, the Chinese government provides tax credit to the projects listed under the Guidance Index of National Renewable Energy and authorizes Chinese State Council to regulate specific rules.[22] So the government could provide tax credit to the development of offshore wind power. But the State Council has no specific provisions so far.

In the "National Offshore Wind Power Development and Construction Program," issued by the National Energy Administration, the fourth chapter requires the grid companies actively involved in offshore wind power development and construction projects supporting the construction of the grid work, the implementation of grid

19. K Liu. Offshore wind power project stranded cost is too high after the start of a loss, available at http://www.china5e.com/news/news-841347-1.html. 2013-07-25 09:18:56 (刘科：海上风电项目搁浅 成本太高启动后肯定亏本，时代周报2013年7月25日).
20. *Notice of the General Office of State Development and Reform Commission and General Office of the Ministry of Finance on declaration of foreign government loan alternative project in 2015([2015]2592),* available at http://bgt.ndrc.gov.cn/zcfb/201510/t20151020_755141.html (accessed by January 27, 2017).
21. *Renewable Energy Law of the People's Republic of China (2009 Amendment)* Art. 24.
22. *Ibid.* Art. 26.

access and consumer market, in time for the grid and supporting documents and arrangements for construction funds to speed up the construction of supporting power transmission projects to ensure the offshore wind power projects and to support the completion of synchronous production grid.[23] According to the third chapter of the "Renewable energy power generation management regulations," the grid companies take the responsibility to access to grid. The whole picture of the grid still is the responsibility of the State Grid.[24]

[3] Incentive for Construction Harbor, Construction Vessel and Grid, or Turbine Manufacturers

According to the fourteenth section of "The Interim Management Measures for the Renewable Energy Special Fund," the special fund for renewable energy development could be used to promote the localized production of renewable energy development and utilization equipment.[25] Since January 1, 2016, with the wind turbine and components imported from abroad, tariff rates significantly reduced, and the acquisition cost of wind power units began to decline. In 2016, the provisional import tariff rate for wind power-driven wind turbines and wind turbine components reduced to 5%, and for wind power equipment to 1%. The rate decline will help reduce the cost of offshore wind farm. So the government could use the fund to invest on the shores and other heavy-duty machines related to the offshore wind power.

[B] Regulations

[1] License Scheme

There are related laws to get the license of the offshore wind farm, such as the "Administrative Licensing Law," "Marine Use Management Law" and "Interim Measures for the approval of enterprise investment projects." Specific to the legal provisions, the most important basis is the "Administrative Measures for the Development and Construction of Offshore Wind Power," developers need to apply to the national marine administrative departments of the sea area with the application documents and maritime use of the argument; after the review, to the State Oceanic Administration departments to obtain pre-sea views of the project issued by the enterprise; and to the national energy authorities to declare the project construction approval application,

23. *Notice of the National Energy Administration for the issuance of national development and construction of offshore wind energy programme (2014–2016)*, available at http://zfxxgk.nea.gov.cn/auto87/201412/t20141212_1869.htm (accessed by February 12, 2017).
24. *Renewable energy power generation management regulations* Chapter 3.
25. *The Interim Management Measures for the Renewable Energy Special Fund* Art. 14.

with the necessary documents issued by the national marine administrative departments of pre-sea observations.[26] Offshore wind project involving land use and maritime planning needs to go through administrative licensing. First, according to the "Catalogue of government approved investment projects," wind power projects need approval.[27] And, as mentioned in the "Interim Measures for the Management of the Early Stage of Wind Power Project," the right to exploit the wind farm construction project is determined through bidding and entrustment.[28] Specific to offshore wind power, "Interim Measures for Management of the Development and Construction of Off-Shore Wind Power" mentioned that without permits, enterprises shall not begin construction of wind plant projects.[29] So, to start the project, the developers have to get the development rights from the state energy authority and the sea-use rights from the State Oceanic Administration authority.

[2] Environmental Impact Assessment

In China, the EIA would run through the preparation, operation and finalization of the whole offshore wind power projects. In accordance with the "*Administrative Measures for the Development and Construction of Offshore Wind Power,*" the developers shall prepare EIA reports based on the requirements and technical standards as embodied in "Marine Environment Protection Law" and "Regulations on Prevention and Control of Marine Environmental Damage by Marine Construction Projects." These reports should be submitted to the State Oceanic Administration for its verification.[30] The National Energy Administration will not approve the offshore wind power project if there is no EIA report or EIA reports have not been approved by the State Oceanic Administration.

There are different contents of the environmental assessment at different stages for the offshore wind project. According to the characteristics of offshore wind power projects, pre-site is about the site selection, if reasonable which can avoid lots of serious environmental problems. After the implementation of the project, the environmental impact of the construction period is mainly due to the impact of ecological environment, the impact on the channel and the impact on the sea bed and so on. The environmental impacts during operation mainly include the influence of sea surface landscape, the influence of electromagnetic radiation and the influence of noise. There is no preferential treatment to facilitate application of offshore wind, which is even harder than onshore wind. Like the Guangdong's first offshore wind power project, investment of CNY 4.2 billion in Zhuhai Guishan offshore wind power project has been stalled for more than a year's time, and the main reason is the project has not passed

26. *Administrative Measures for the Development and Construction of Offshore Wind Power Chapter 3.*
27. *Catalogue of government approved investment projects (2016)* Art. 2.
28. *Interim Measures for the Management of the Early Stage of Wind Power Project* Art. 13.
29. *Interim Measures for Management of the Development and Construction of Off-Shore Wind Power* Arts 21/22.
30. *Administrative Measures for the Development and Construction of Offshore Wind Power* Arts 24/25.

through the Marine EIA.[31] And on September 8, 2016, the project finally started. In practice, it is a long process for offshore wind power enterprises to pass the EIA. On the one hand, there are not uniform technical standards of the EIA, and there is lack of coordination between different governmental units and between central and local governments. On the other hand, the level of regulations on offshore wind power is comparatively low which has negatively influenced the implementation of these regulations.

[3] Planning the Legal Regime

The most relevant laws about planning the legal regime of the offshore wind farm include the "Administration of the Use of Sea Areas," "Marine Environment Protection Law," "Administrative Measures for the Development and Construction of Offshore Wind Power" and "Implementation Guide on the Interim Measures for Management of the Development and Construction of Off-Shore Wind Power."

Offshore wind power planning includes national offshore wind power plans and offshore wind power plans of coastal provinces. The roles of offshore wind power land under the spatial planning and/or marine space planning. The preparation and management of national offshore wind power development planning is organized by the state energy authority which uniformly and together with the State Oceanic Administration authority, reviews and approves the offshore wind power plans of coastal provinces. For offshore wind power construction projects, tendering method is prioritized to select investment enterprises for development. For the expansion of offshore wind power projects, the original project entity may submit application and, after approval by the state energy authority, obtain development rights for project expansion.

Article 6 of the "Administrative Measures for the Development and Construction of Offshore Wind Power" provides that the planning of offshore wind power projects should be consistent with national renewable energy marine functional zonings and development planning of ocean economy.[32] Article 7 of this Regulation also asks about when making the offshore wind farm, the whole plan needs to be considered in accordance with the requirements of ecological civilization construction, the overall consideration of the development of strength and resource environment carrying capacity.[33] Article 19 of this Regulation provides that the construction sea use of

31. Offshore wind power predicament survey: South Network 4.2 billion project stuck environmental impact assessment: As early as April 16, 2013, for the project to carry out environmental impact assessment of the Shanghai Survey and Design Institute on the release of the EIA publicity. The EIA shows that the high frequency noise generated by the construction of the fan pile foundation in the Guishan project in Zhuhai will affect the hearing of the Chinese white dolphin, which causes its foraging and social activities to be disturbed. Water pollution through the food chain will affect the health of Chinese white dolphins, but also easy to make the Chinese white dolphin skin infection. But the report argues that these effects are temporary and reversible. Available at http://money.163.com/15/1017/08/B647230Q00253B0H.html (accessed by February 14, 2017).
32. *Administrative Measures for the Development and Construction of Offshore Wind Power* Art. 6.
33. *Ibid.* Art. 7.

offshore wind project should comply with the principles of saving and intensive utilization of marine resources, and should be reasonably distributed.[34]

Article 5 of the "Implementation Guide on the Interim Measures for Management of the Development and Construction of Offshore Wind Power" provides that the marine space planning of offshore wind projects should be consistent with the national renewable energy development planning, and should comply with the marine functional zonings, island protection planning and marine environmental protection planning. The location of offshore wind power projects should be more than 10 km away from the coast, and the sea should be at least 10 m deep when the breadth of the shoals is wider than 10 km. This project should not be located in various marine natural conservation zones, marine special preservation park, important fisheries waters, and so on.[35]

Reasonable choice of infrastructure is one of the main considerations. When the government starts their planning for the wind power, they should first communicate with the relevant authorities to determine the scope of these key sensitive areas and the required protection distance, and then adjust the scope of wind farm planning to avoid key sensitive areas and the required protection. The layout of offshore wind farms shall not be planned in all kinds of marine nature reserves, special marine protected areas, important fishery waters, typical marine ecosystems, estuaries, gulfs, natural historical relics protection areas and other sensitive sea areas.

[4] International Law

The offshore wind farm projects need to occupy a certain area of the marine space, this will produce a series of problems, such as: the right to use the sea; marine renewable energy facilities or structure of the legal status and security issues; the impact of facilities and structures on maritime traffic safety; the impact on the marine environment; and even the security aspects of the defense. The development and utilization of oceans and seas should be carried out under a series of international legal frameworks based on the 1982 United Nations Convention on the Law of the Sea. And according to the "Implementation Guide on the Interim Measures for Management of the Development and Construction of Offshore Wind Power," offshore wind farms in principle should be in the offshore distance of not less than 10 km, the beach width of more than 10 km when the water depth is not less than 10 m of the sea area layout.[36] It makes the wind farms move into the deep sea, which brings difficulties for the offshore wind farms construction. There are still no such problems, mainly because that kind of big offshore wind farms construction has not shaped in Asia.

34. *Ibid.* Art. 19.
35. *Implementation Guide on the Interim Measures for Management of the Development and Construction of Offshore Wind Power* Art. 5.
36. *Ibid.*

[5] Grid

In China, the State grid and the five regional power grid enterprises are responsible for the grid system on land; there is no power grid at sea. Traditionally offshore oil and gas field were developed as a single group of power stations aimed at power generation; the current development of the regional power grid is to achieve the maritime power transport. The State Oceanic Administration is responsible for the laying of submarine cables, pipelines and routing surveys, surveys and other related activities for the laying of the sea, the territorial sea and the continental shelf of the People's Republic of China. According to "Provisions on the Administration of Laying Submarine Cable Pipelines," submarine cables and pipelines owners shall apply the application to the main pipe authority sixty days prior to the implementation of the routing survey for the pavement, and the approval should be obtained sixty days before the laying of the plan, the final determination of submarine cables, pipeline routing.[37] And the routes for the submarine cables beyond the oil development zones laid by the offshore oil development shall be reported to the competent authority prior to the approval of the overall development plan for the oil and gas fields and shall be subject to the approval of the competent national competent energy authority. Offshore wind farms suit the same regulation.

In this way, offshore wind developers need to apply the application to the main pipe authority, and report to the competent authority. We can see from the license, planning, and the laying submarine cable pipelines, the developers need to ask for the approval again and again, which could discourage the developers.

[6] Other Concerns Fisheries

Offshore wind farms could be generally traditional fishing areas. Therefore, the construction of offshore wind farm is bound to occupy some of the traditional fishing sea area, if the spacing between the offshore wind farms is too small, it will affect the fishing vessels, fishing activities and offshore fishing space, a direct impact on the interests of fishermen which is one of the factors constraining the development of offshore wind power. And according to the "Administrative Measures for the Development and Construction of Offshore Wind Power," offshore wind farms in principle should be in the offshore distance of not less than 10 km, the beach width of more than 10 km and sea depth of not less than 10 m of the sea area layout.[38] At the same time, layout of offshore wind farms shall not be planned in any kind of marine nature reserves, marine special protection areas, important fishery waters, typical marine ecosystems, estuaries, gulfs and natural historical relics protection areas. This will increase the difficulty of offshore wind power development. As the Donghai intertidal wind farm, a 200,000 KW project in the sea site and Jiangsu coastal beach reclamation area, and Yancheng rare birds nature reserve overlapped, Donghai project adjusted its

37. *Provisions on the Administration of Laying Submarine Cable Pipelines* Art. 5.
38. *Administrative Measures for the Development and Construction of Offshore Wind Power* Art. 7.

position in the sea.[39] This issue was not pointed out by the fishermen, but by the Bureau of the ocean; there is no direct conflict between the fishermen and the offshore wind farms.

§8.04 CHALLENGES AND SOLUTIONS

[A] Institutional Design

Offshore wind farms project is a kind of new energy in China, which is still in the early stage of development. The promotion of the offshore wind farms is basically done by the policy and the authorities. So it is very important to confirm the responsibilities of each part, and make sure that policies are playing a positive role.

To speed up the introduction of renewable energy quota system, a clear local government, power grid enterprises and power generation business responsibilities form the mechanism to promote the completion of wind power development goals. And we need to continue to increase the wind power industry policy support to ensure that wind power meet normal and reasonable level of profitability, and enhance industrial investment and development confidence.

[B] Incentives

As a new industry, offshore wind farms are not competitive like other power industries, even onshore wind power is much more competitive. The offshore wind power unit cost is higher than the onshore wind power, the main cost being offshore engineering. The offshore wind power development has gained experience from the pilot project, pilot projects, large-scale installed capacity of the process, in which the costs are reducing. The most effective incentive is building a competitive market. But there are still a lot of preparation for the offshore wind to build a competitive market. So it needs more incentives for the beginning stage.

Considering the complementary policy, we should use different policy tools for different phases: in the R&D stage more efforts to increase public finance for research activities, in the commercial demonstration more efforts to increase finance investment, in the growth stage more government procurement to protect the investment, in the popularization stage more tax incentives to build competitive markets.

[C] Regulations

In the field of offshore wind power, the main purpose of the law is to guide the role of the actual work of the policy. This will make the policy as mainly problem-oriented, and bring about changes to the local and central government, where the same level of

39. The second batch of tenders will start the first batch of offshore wind power concession project plight to be solved, available at http://news.bjx.com.cn/html/20110713/294884.shtml (accessed by January 13, 2017).

decree is handled inconsistently by the different regulatory authorities. It the end, it will weaken the implementation of policy.

In order to achieve the guiding role of the law, we need to further clarify the responsibilities of different departments, establish a clear approval process and improve administrative efficiency.

§8.05 CONCLUSION

Offshore wind power is a new renewable energy industry in China. The development of offshore wind power is of great significance to promote the adjustment and optimization of energy structure and the transformation of economic development mode in coastal areas. In recent years, as the state introduced a series of policy guidance and encouragement, China's offshore wind power started. Renewable energy development "Twelfth Five-Year Plan" mentioned that by 2015, China's offshore wind power installed capacity must be of 500 million KW, which is the basic form of a complete, internationally competitive wind power equipment manufacturing industry.[40] However, according to the National Energy Administration data, as of the end of 2015, China has completed the installed capacity of offshore wind power project total 1014.68 MW.[41] Now, the "Thirteen Five-Year Plan" is about to begin. How to break the current offshore wind power development dilemma is now a challenge. To solve this problem, we need to combine the existing system, to find out a way to overcome the difficulties. From the institutional design, incentives and regulations, we could see the problem and identify the solutions in the future.

40. *Renewable energy development "Twelve-Five" plan* (August 6, 2010), available at http://news.bjx.com.cn/html/20120810/379617-5.shtml (accessed by January 18, 2017).
41. China Wind Energy Association Professional Committee of CWEA. (2015). *Summary of China's wind power installed capacity statistics in 2015 (1 Jan. 2015 to 31 Dec. 2015)*. Available at http://www.cnenergy.org/xny_183/fd/201604/t20160405_276532.html (accessed by January 15, 2017).

CHAPTER 9

Legal Framework to Develop Offshore Wind Power in Korea

Eubong Lee

§9.01 INTRODUCTION

According to *the Framework Act on Low Carbon, Green Growth* of Korea that was enacted in 2010, the Government shall establish a master plan for energy every five years (hereafter referred to as "master energy plan" in this chapter) for a planning period of twenty years in accordance with basic principles for policies on energy.[1]

Although South Korea had set up "green growth" as the main national development policy goal since enactment of *the Green Growth Act*, the portion of renewable energy including wind power remains small part of whole energy generation sources with 225 MW of new wind power generation installations in 2015, and total installed capacity just over 835 MW (*see* Figure 9.1).[2]

In 2014, the Korean government pronounced the second master energy plan[3] and the fourth long-term master plan on the new and renewable energies,[4] which implies the planned rate of each energy sources including new and renewable energies. These

1. *Framework Act on Low Carbon, Green Growth*, Act No, 11965, July 30, 2013, Art. 41.1.
2. Taiwan added 14 MW of new capacity in 2015, bringing its total installed capacity to 647 MW. Global Wind Energy Council, *Global Wind Report Annual Market Update 2015*, Global Wind Energy Council, p. 10, *available at* http://www.gwec.net/wp-content/uploads/vip/GWEC-Global-Wind-2015-Report_April-2016_19_04.pdf.
3. *The 2nd Korea Energy Master Plan, Outlook & Policies to 2035*, The Ministry of Trade, Industry and Energy of Korea, January 2014, *available at* http://www.motie.go.kr/motie/py/td/tradefta/bbs/bbsView.do?bbs_cd_n=72&cate_n=3&bbs_seq_n=104479.
4. *The 4th Long-term Basic Plan on the New and Renewable Energies*, The Ministry of Trade, Industry and Energy of Korea, September 9, 2014, *available at* http://www.motie.go.kr/motie/ne/rt/press/bbs/bbsView.do?bbs_seq_n=79321&bbs_cd_n=16.

plans purport to increase the rate of new and renewable energies supply from 3.6% in 2014 up to 11% by 2035 with an annual growth rate of 6.2%.[5]

At the end of 2015, "The Annual Report of wind-power association" reports that wind energy covers 7.6% of the whole energy facilities in Korea, which accounts for up to 11.24% of the new and renewable energy sources. In terms of the rate of power generation, the rate of wind power generation decreased to 0.27% of the whole energy sources, which in turn accounts for 6.48% of the new and renewable energy sources. Nevertheless, the onshore and offshore winds are planned to account for 18.2% of the whole new and renewable energy supply by 2035 according to the national master plans above. (*See* Table 9.1)

Table 9.1 The Planned Rate of New and Renewable Energy Sources Supply by the Long-Term Master Plan on the New and Renewable Energies (2014)[6]

	2012	*2014*	*2025*	*2035*	*Annual Growth Rate*
Solar-thermal	0.3	0.5	3.7	7.9	21.2
Solar-PV	2.7	4.9	12.9	14.1	11.7
Wind	**2.2**	**2.6**	**15.6**	**18.2**	**16.5**
Biomass	5.2	13.3	19.0	18.0	7.7
Hydraulic	9.3	9.7	4.1	8.5	18.0
Geothermal	0.7	0.9	4.4	8.5	18.0
Marine	1.1	1.1	1.6	1.3	6.7
Waste	68.4	67.0	38.8	29.2	2.0

Along with the strong government-driven policy, the green growth strategy, originally the Korean government planned to enlarge the offshore wind power generation, with consideration of limitation of generation capacity and possible claims from the regions near onshore wind generations sites.

In 2010, the Korean government pronounced "The Roadmap for Offshore Wind Policy" which is composed of three steps each of which is from the test-beds construction (2011–2013), the test farms construction (2014–2016) and dispersion of wind farms (2017–2019) from 2011 to 2019. (*See* Table 9.2)

5. However, the Basic Energy plan expects that, until 2035, the rate of nuclear energy will increase up to 27.2% from 19.0% in 2011, while that of the coal and oil will decrease from 65.8% to 52.0%.
6. Jaehee Park, *Report on Wind Power Industry 2015, South Korea*, March 1, 2016, Netherlands Enterprise Agency (RVO.nl), pp. 1-2, *see* Table 2, *available at* https://www.rvo.nl/sites/default/files/2016/03/Rapport%20Windenergie%20Zuid-Korea.pdf.

The target capacity of the offshore wind power generation was 900 MW up to 2016 and 1.5 GW, until 2019. The main project to achieve this goal was the large-scale project, the 2.5 GW wind power farm project in the south-western sea. (*See* Figure 9.2)

Table 9.2 The Roadmap for Offshore Wind Policy (2010)

	Test-Beds Construction	*Test Farms Construction*	*Dispersion of Wind Farms*
Period	2011–2013	2014–2016	2017–2019
Object	• Establish the test bed • Secure the track records • Develop the core technologies	• Secure the technologies for operation • Test feasibility of business	• Lower cost • Develop the large scale wind farm • Commercial operation
Target capacity	100 MW(5 MWx20 towers)	900 MW(5 MWx180 towers)	1.5 GW(5 MWx300 towers)
Financial resource	R&D, projects(603.6 bil. KW)	R&D, projects(30,254 bil. KW)	Projects(56,300 bil. KW)
Developer	public, private	public, private	Private

In 2011, the Korean government allocated 135 hundred million KW to the promotion of the new and renewable energy industries, which was more than the previous year by 24.1%. Recently, in July 2016, the Korean government pronounced that it promotes the development of the new energy industries including new and renewable energies more aggressively with investment of 42 trillion KW up to 2020 and plans to newly construct more than three offshore wind farms in Taean, Jeju and Gori.

Regardless of the suggested goals and plans above, it seems that there is still a long way to go for Korea. For example, as the planned progress of the south-western offshore wind farm project was delayed owing to the regional unfavorable opinions and tendering companies' hesitation,[7] the original development plan was adjusted and downsized recently.[8]

Based on these facts, from the next section, this chapter will deal with more concretely on what kinds of institutions, programs and norms Korea has designed to promote the offshore wind power. Also, it would include the hurdles in arriving at its goal and solutions to overcome it.

7. Even though, in the early stage, eight wind turbine manufacturers including Hyundai Heavy Industries, Samsung Heavy Industries, Doosan Heavy Industries and Hyosung Heavy Industries had shown intention to participate, the rest except Doosan have retreated from the projects. However, Hyosung decided to rejoin recently.
8. Although the planned capacity in the Phase I, when the test beds are supposed to be established, was 99.5 MW in 2011, it was revised down to 60 MW in 2016. Also, the period of the Phase I and II was extended by two years each.

Figure 9.1 Offshore Wind Capacity in 2015[9]

GLOBAL CUMULATIVE OFFSHORE WIND CAPACITY IN 2015

5,000 MW
4,000
3,000
2,000
1,000
0

Cumulative Capacity 2014
Cumulative Capacity 2015

ANNUAL CUMULATIVE CAPACITY (2011-2015)

14,000 MW
12,000
10,000
8,000
6,000
4,000
2,000
0

4,117 (2011)
5,415 (2012)
7,046 (2013)
8,724 (2014)
12,107 (2015)

	UK	Germany	Denmark	PR China	Belgium	Netherlands	Sweden	Japan	Finland	Ireland	S Korea	Spain	Norway	Portugal	US	Total
Total 2014	4,500.4	1,012	1,271	654	712	247	212	50	26	25	5	5	2	2	0.02	**8,724**
New 2015	572.1	2,282.4	0	360.5	0	180	0	3	0	0	0	0	0	0	0	**3,398**
Total 2015	5,066.5	3,294.6	1,271.3	1,014.7	712.2	426.8	201.7	53	26.3	25.2	5	5	2.3	2	0.02	**12,107**

Source: GWEC, 2016

Figure 9.2 The Location of the South-West Offshore Wind Farm

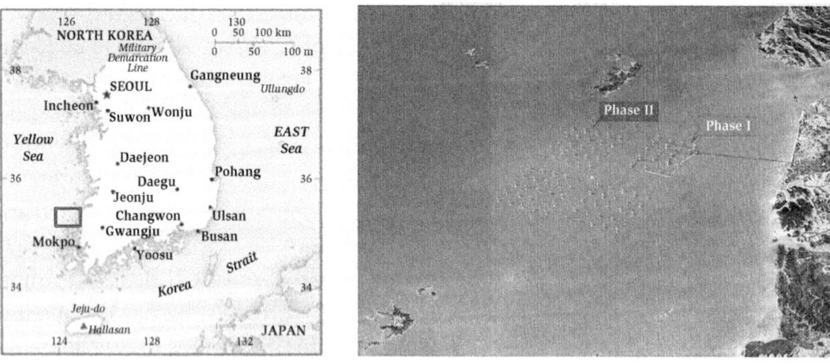

§9.02 INSTITUTIONAL DESIGN FOR OFFSHORE WIND

[A] The Authorities for Development of Offshore Wind Power

In Korea, in order to develop the power generation business, operators shall get permits with required conditions met.[10] In term of electricity generation permit, for the new and renewable energy projects over 3 MW, including wind energy, the electricity provider shall get it from the Ministry of Industry, Trade and Resources, while for those below 3 MW, from the Metropolitan Governments or the provincial governments.

Exceptionally, for Jeju Island, the Governor of Jeju has the authority to give a permit to the energy facilities with the generation capacity of less than 20 MW. Recently, the Governor of Jeju revoked the license for the business of one onshore wind farm since it was found that the manager of the energy company had given bribery to the representative of the community cooperated-farm.[11]

Besides the license for the power generation business, the enterpriser shall get the permit for occupation, utilization and reclamation of water surface from the authorities managing the public water surface. The authority managing the exclusive economic zones and the harbor zones is the Minister of Oceans and Fisheries, and those of other areas are the governors of metropolitan and the provincial governments, and basic level local governments. Since in marine areas the boundaries of administered area are often unclear, conflicts over use of waters among local areas often arise.

For more effective promotion of the electric power projects, *the Electric Source Development Promotion Act* allows substitution of the various kinds of permits under laws including the permit for occupation and use of public water surface with the approval of the Minister of industry, trade and resources for the electric source development action plan.[12] However, the approval cannot replace the environmental impact assessment and the disaster impact assessment.

When it comes to the R&D projects and subsidies, *the Korea Energy Agency* founded under the Ministry of Industry, Trade and Resources is responsible for the enforcement, administration and allocation of funds. For managing practical affairs on renewables, the Korea Energy Agency also set up the Center of new and renewable energy.

For enforcement and promotion of offshore wind policy, in 2010, the Korean government established "The Offshore Power Promotion Taskforce" under the Ministry of Knowledge Economy, the former ministry of the Ministry of Trade, Industry and Energy. In 2012, to develop "the South-Western Sea Offshore Wind Power Project," the government set up the special purpose company (SPC), "the Korea Offshore Wind Power" which was invested by six power generation companies.

9. Global Wind Energy Council, *see supra* note 2, p. 49.
10. *Electric Utility Act*, Act No. 13858, January 27, 2016, Art. 7.
11. Yonhap News Agency, "The first revocation case of the Jeju wind-farm permit," (July 26, 2016), http://www.yonhapnews.co.kr/bulletin/2016/07/26/0200000000AKR20160726175600056.HTML?input=1195m (last visited August 1, 2016).
12. *Electric Source Development Promotion Act*, Act No. 13805, January 19, 2016, Art. 6. For example, after the three-year delay of the South-West Wind Farm project, where it was hard to get the permit for occupation and use of public water from the relevant local governments, the Minister of industry, trade and resources have finally issued the approval of the electric source development action plan in 2016.

[B] The Role of "Law" under the Institution Decision Making and Related Policies

The development of offshore wind energy is deeply related to the CO_2 emission reduction policy. Thus, the energy regulations, environmental regulations and economic and industrial regulations regarding CO_2 emission reduction affect the future of offshore wind energy.

One report by GWE in 2014[13] tells that we will have a different future depending on which strategy we will take: first, the advanced scenario which is based on the assumption that the nation will take the most ambitious policy with the best goal: second, the moderate scenario which is based on the assumption that the nation will keep pursuing the energy and emission reduction policy to meet the targets suggested in Cancun and achieve them: the new policy scenario (similar to the current policy scenario) which is based on the current directions and intentions, but not backed up by law.

This report compared the CO_2 emission reduction, the growth of wind powers, the cost for production, investment and job creation at the current time, 2013 or 2014 with the expected results in the future like 2020 or 2030. (*See* Table 9.3) At least, under the new policy scenario that is not supported by legal enforcement, the wind power industries are expected to not grow much compared to the current situations.

13. Global Wind Energy Council & Greenpeace International, *Global Wind Energy Outlook 2014*, GWEC (2015). pp. 56–57, *available at* http://gwec.net/wp-content/uploads/2014/10/GWEO2014_WEB.pdf.

Table 9.3 The Prospect of Wind Power Industries Based on Each Scenario of GWEC

Scenario	*Year*	*Wind Power Production (TWh), (%))*[14]	*Global Annual Growth Rate (%)*	*Capital Cost (EUR/kW)*	*Investment (EUR 1,000)*	*Job Totals (Person)*	*CO2 Reduction (with 600 g CO2/kWh) [Annual Mio t CO2]*
Advanced scenario	2013	620 (2.9%)	15% (2014)	1,258	44,606,048	601,519	372
	2020	1,850 TWh (8.1%–8.8%)	13%	1,137	97,920,604	1,450,753	1,178
	2030	5,000 TWh (16.8%–18.9%)	7%	1,100	129,261,055	2,171,804	3,050
Moderate scenario	2013	620 (2.9%)	14% (2014)	1,258	44,606,048	601,519	372
	2020	1,750 TWh (7.2%–7.8%)	10%	1,214	79,856,761	1,090,378	1,048
	2030	3,900 TWh (12.9%–14.5%)	6%	1,203	101,864,571	1,504,698	2,333
New policy scenario	2013	620 (2.9%)	12% (2014)	1,252	44,606,048	601,519	372
	2020	1,500 TWh (6.2%–6.7%)	6%	1,405	48,225,762	635,439	844
	2030	2,535 TWh (8.4%–9.4%)	3%	1,418	41,875,598	644,815	1,387

14. Wind power penetration of world's electricity.

As far as in Korea, surely the law has certain limitations to promote consistently the offshore wind power industries, while they cannot be ignored in that they provide the framework and systemized capacities for the wind power generations to feasibly work.

Although the law provides much policy tools to promote the renewables, realization of such policy mostly depends on budgets which are decided by how much weight the government places on renewables. Since the companies participating in the projects hardly get benefits, reluctant are they to invest in new business without constantly assured policy promotion.

Another big hurdle is the legal structure that have already been constructed disregarding the features and characteristic of the new way of power generations or weighing the traditional energy resources more which cover large portion of the whole energy production.

Also, there are lots of areas to be added in energy laws to make the offshore wind development feasible. Over offshore wind energy development as well as the renewable energies, the new legal questions arise; the new types' rights and obligations and how to use the natural resources among diverse interested parties.

§9.03 LEGAL DESIGN FOR OFFSHORE WIND

[A] Incentives

[1] Main Finance Scheme

The Korea government has adopted the Renewable energy portfolio standard (RPS) since 2010, while it had executed FIT until then. During the FIT-adopted period from 2000 to 2008, wind energy industry has grown by thirty times in Korea. After shifting to RPS, the wind power industries used to say that the RPS is not enough to lead to constant progress so the FIT is still needed.

Regarding the Renewable energy portfolio standard (RPS), *Act on the Promotion of the Development, Use and Diffusion of New and Renewable Energy Act*[15] requires the electric energy producers who own non new and renewable facilities with more than 500 MW installed capacity[16] to supply at least a certain amount of electricity generated by using new and renewable energy. Such mandatory amounts are decided by the Ministry of Industry, Trade and Resources every year with the rate of less than 10% of the whole amount of electric generation supplied by the producers required. This RPS rate of 2016 is 3.5%, which increase by 0.5% until 2019, by 1.0% every year from then. (*See* Table 9.4)

15. *Act on the Promotion of the Development, Use and Diffusion of New and Renewable Energy*, Act No. 13087, January 28, 2015, Art. 12-5.
16. The operators included this condition are sixteen power generation companies.

Table 9.4 The Renewable Energy Portfolio Standard (RPS) by Year

Year	2012	2013	2014	2015	2016	2017	2018	2019	2020	2021	2022	2023	After 2022
RPS Rate (%)	2.0	2.5	3.0	3.0	3.5	4.0	4.5	5.0	6.0	7.0	8.0	9.0	10.0

Actually, these rates were lower than the original target. Yet, recently again it was planned to be modified to be higher. According to the new plan on the New Energy Business Projects pronounced by the Minister of Industry, Trade and Resources released in July 5, 2016, from 2018, the mandatory rate for energy producers to provide new& renewable energy under RPS will go up to 5.0% in 2018, 6.0% in 2019 and 7.0% in 2020.

If the electric operators fail to meet the obligation over the legal rate of RPS, they should be charged monetary penalties or alternatively may purchase a new and renewable energy supply certificate (REC) and appropriate it for mandatory supply.[17] To weigh the values differently and reflect policy purpose on each renewable energy resource in REC, regulations set the different REC Multiplier for each renewable energy. REC Multipliers of the offshore wind power and the Energy Storage System (ESS) facilities connected with offshore wind facilities are relatively high compared to onshore. (*See* Table 9.5)

Also, in order to increase and extend the use or distribution of new and renewable energy, the Minister of Trade, Industry and Energy can require the State and a local governments, public institutions, government-contributed institutions, and the governments and public institutions-invested corporations to mandatorily install new and renewable energy facilities in newly built, extended, or remodeled buildings in order to use energy supplied from new or renewable energy over a certain percentage of the estimated amount of energy.[18]

Table 9.5 REC Multiplier by Types of Renewable Energy

<table>
<tr><th>REC Multiplier</th><th colspan="2">Types of Energy</th></tr>
<tr><td>0.7 ~ 1.5</td><td colspan="2">Solar energy</td></tr>
<tr><td>1.0</td><td colspan="2">Bio-energy</td></tr>
<tr><td>1.0</td><td colspan="2">Hydroelectric power</td></tr>
<tr><td>1.0</td><td colspan="2">Onshore wind power</td></tr>
<tr><td>1.5</td><td colspan="2">Offshore wind power (below 5 km)</td></tr>
<tr><td>2.0</td><td rowspan="2">Offshore wind power (over 5 km)</td><td>Fixed</td></tr>
<tr><td>1.0 ~ 2.5</td><td>Floating</td></tr>
</table>

17. Act No. 13087, *See supra* note 15, Art. 12-5.
18. Act No. 13087, *See supra* note 15, Art. 12.

REC Multiplier	*Types of Energy*	
5.5	ESS facilities (connected with offshore facilities)	2015
5.0		2016
4.5		2017

[2] Supplementary Finance Scheme

[a] Low Interest Loan and Loan Guarantee

The Government shall formulate financial support, taxation or other necessary support measures, for a person who runs a business on new and renewable energy or is recommended by the Government to invest or have duties on RPS, a person engaged in the technological development, use, and distribution of new and renewable energy, or a person whose facilities have been certified.[19]

For this financial support, the Korean Government provides special loans with low interests and long terms for production, operation and facilities on new and renewable energy through the government-invested banks or private banks. For the period of a year beginning 2016, the government funds 100 billion KW for these special loans. If the business owner wants to receive such a loan from such banks, it needs to get a recommendation letter from the new and renewable energy center which is required to submit to the bank for the review.

[b] Tax Credit

Based on *the Restriction of Special Taxation Act*,[20] when a legal person or a private person invests in the energy saving facilities including new and renewable energy, he or she can get tax deduction for the certain portion of the amount of the investment. The deduction rates are 6% for MSEs, 3% for mid-sized enterprises.

[c] Investment Subsidy/Grant

If the home owner installs new and renewable energy facilities on her or his own house, the government can subsidize a certain portion of the cost for installation. Also, the owners of buildings excluding those belonging to the state and local governments are eligible to get a certain portion of the cost for installation of the new and renewable energy facilities.

Furthermore, the local governments afford to receive subsidies from the central government for their projects for the purpose of improving the conditions of demand

19. *Id.*, Art. 29.
20. *The Restriction of Special Taxation Act*, Act No. 12173, January 1, 2014, Art. 10.

and supply on energy and develop the local economy through distribution of the environmentally friendly new and renewable energy fitting the regional characteristics.

[d] Community Support

Korean legislations provide special concerns on regions affected by energy facilities. *The Act on Support to the Power Plants-Neighboring Areas*[21] defines "the power plants-neighboring areas" as Eup, Meyun, Dong (the basic level of local unit) including lands and islands within 5 km from power plants. When it comes to the new and renewable energy projects, the power plants have to have over 2,000 kW of power generation capacities. Yet, this law has several exceptions on such definitions to support the area outside of the Neighboring Areas: for the effective enforcement of projects and balanced development of regions; the area outside of the Neighboring Areas of the nuclear power plants; the mass mitigation area due to the power plants construction.

The financial sources for support comes from the Fund of Electric Industries' Foundations founded by:[22] the charges or additional charges of electricity users (3.7% of electricity's price, 75.25% of the fund's income); the penalties on violation of the duty of Renewable energy portfolio standard (RPS); profits from the fund operation; and engineering fees from R&D funded by The Fund of Electric Industries' Foundations.

The types of support projects include the basic support projects, the special support projects, the public relations projects, as well as other projects, and the direct support by project operators.[23] The basic projects use annual funding to support the income growth programs, the public welfare program, the residents' welfare support program, the company recruitment program and the electric fee subsidy by annual funds. The programs of special support projects are decided by the governors of local areas. Yet, these projects can be allocated by the Minister of Industry, Trade and Resources with an amount less than 15% of the costs of power plant construction through the review of the committee.[24] Also, the Minister of Industry, Trade and Resources can allocate less than 10% of the costs of power plant construction through the review of committee, for other support purposes. The operator's direct support funds are exclusively for the nuclear power plants and the water power plants.[25]

---

21. *The Act on Support to the Power Plants-Neighboring Areas*, Act No. 13151, February 2, 2015, Art. 2.
22. *Id.*, Art. 13.
23. *Id.*, Art. 10.
24. For the nuclear power plants and thermal power plants (flaming coal), there are the additional special support fund with the amount of 0.5% of the costs of power plant construction.
25. The support amounts for nuclear power plants are the amount of power generation two years before (KWh) × 0.25 (KW / KW h); those of water power plants are 5,000 thousand KW per the installed capacity.

This law also provides the rules on how to allocate the funds among two or more Power Plants Neighboring Areas.[26] In principle, the funds are allocated with consideration of the proportion of territories, the populations, distance from power plants and the location. Based on each factor for consideration, the rate for allocation is different by 40%, 30%, 20% and 10% as shown in Table 9.6.

Table 9.6 The Rate for Fund Allocation[27]

Allocation Rate	*Factors for Allocation*
40% of funds	By the proportion of the area of each Eup, Meyun, Dong (the basic level of local unit) including lands and islands within 5 km from power plants (excluding the hydroelectric power plants and the tidal power plants)
30% of funds	By the proportion of the population in the area of each Eup, Meyun, Dong (the basic level of local unit) including lands and islands within 5 km from power plants (excluding the hydroelectric power plants and the tidal power plants)
20% of funds	For the area of each Eup, Meyun, Dong (the basic level of local unit) where the power plants are located (for two or more plants, by the number of generators)
10% of funds	By allocation of the Minister of Industry, Trade and Resources with 10% of funds after the review of the regional committee considering the level of regional development, the balance of allocation among the areas and the regional conditions

Nevertheless, there are exceptions based on the characteristics of regions; if the principle is strictly followed, then it would cause problems in construction and operation of power generation plants as corresponding to one of the three conditions; first, the power plant is located in a populated area such as Dong (the basic level of the administrative unit of the city) of the Metropolitan Governments; second, the conflict among municipalities is severe due to the broad surrounding area which have the hydroelectric power plants or the tidal power plants; third, the allocated fund is too small and the power generation operation capacity is less than 10 million KW.

The basic support fund is estimated by the formula as below:[28]

the amount of power generation two years before (KW h) x the unit price of support fund by each power generation source (KW / KWh) x the installed capacity (MW) x the unit price of the installed capacity by each power generation source (10,000 KW/ MW).

26. Act No. 13151, *see supra* note 21, Art. 14-2.
27. *The Presidential Decree on the Act on Support to the Power Plants-Neighboring Areas*, Decree No. 26221, May 1, 2015, Art. 29.
28. Decree No. 26221, *see supra* note 27, Art. 27(1), Table 2; Decree No. 26221, *see supra* note 27, Art. 29.

The unit price of support fund for the new and renewable energy is lower than those of nuclear or thermal energy since the harmful impacts of the new and renewable energy are less than nuclear or thermal energy. However, the offshore wind farm operators ask the government to increase the unit price of support fund in order to give stronger incentives to the regions. The unit price of support fund is provided in the Table 9.7.

Table 9.7 The Unit Price of Support Fund and the Installed Capacity by Power Generation Sources

Power Generation Source	Nuclear	Thermal (Bituminous coal)	Thermal (Anthracite)	Thermal (Oil)	Thermal (Gas)	Pumped Stored	Hydro-electric	Tidal	the new& renewable energy
the unit price of support fund (KW / KW h)[1]	0.25	0.15	0.3	0.15	0.1	0.2	0.2	0.2	0.1
the unit price of the installed capacity (10,000 KW / MW)[2]	-	-	-	-	-	50	500	-	-

1. "KW" stands for the currency Korean Won. The unit for the amount of electricity generation is "kWh" (kw per hour).
2. "KW" stands for the currency Korean Won, a price unit for support fund here. MW here is the unit for the electric facilities' capacity for generation.

[B] Regulations

[1] License Scheme

Any person who intends to conduct electricity business shall obtain permission from the government for each type of electric business.[29] For the new and renewable energy projects, the projects exceeding 3 MW including wind power must be approved by the Ministry of Industry, Trade and Resources, and for those less than 3 MW from the Metropolitan Governments or the Local governments.[30] Exceptionally, in the case of Jeju Island, the Governor has permission to operate an energy facility with a generating capacity of less than 20 MW.

[2] Environmental Impact Assessment (EIA) and Strategic Environmental Assessment (SEA)

The Korean Environmental Impact Assessment Act[31] articulates on three kinds of EIA programs: the strategic environment assessment for planning; the EIA for permits of enforcement plans; and the small scale impact assessment for an area which needs special protection.

When it comes to the strategic environment assessment, when the Minister of Industry, Trade and Resources designates the candidate areas for the electric source development under *Electric Source Development Promotion Act*, it is required to have a presentation meeting for residents and a hearing meeting by the residents' demands in order to collect community opinions.[32]

Regarding the EIA, the power business licensee shall collect community opinions before drafting the EIA for the projects implying the wind power generation facilities and related equipment of which the power generation capabilities are over 100,000 KW.[33] However, there are no special treatments for the wind power generation facilities compared to other energy generation facilities.

If the locations of energy generation facilities are in areas which do not need the EIA, but need special care, a small scale impact assessment is required: the development projects over certain scales within the area for protection and management; the area for the protection of ecosystem and landscape; and the area for protection of wetlands.[34]

29. Act No. 13858, *See supra* note 10, Art. 7.
30. *The Executive Order on the Electric Utility Act*, the Ministry of Industry, Trade and Resources' Order No. 203, July 28, 2016, Art. 4.
31. *The Environmental Impact Assessment Act*, Act No. 14232, May 29, 2016, Art. 7.
32. *Id.*, Art. 13.
33. *Id.*, Art. 25.
34. *Id.*, Arts 43-44.

[3] Site Related Planning

Also, *The Marine Environment Management Act*[35] regulates the consultation on utilization of the coastal area and the impact assessment of the coastal utilization, which are required to go through when constructing offshore wind farms.

The marine management authorities are the Metropolitan Governments, the provincial governments and the Minister of Oceans and Fisheries. The authorities which give the permits for the projects, which are required to get the permit of occupation and use and the certificate of reclamation of public water surface, should go through a consultation with the Minister of Oceans and Fisheries on utilization of the coastal area.[36] When those authorities consult with the Minister of Oceans and Fisheries, those authorities shall ask for the review of the impact assessment of the coastal utilization to the Minister of Oceans and Fisheries with the drafts provided by the project operators.[37] When drafting the assessment papers, the project operators shall hold a presentation meeting, a hearing meeting in order to collect community opinion.[38]

[4] Marine Planning and Land Planning

The Public Waters Management and Reclamation Act[39] articulate on the permit for occupation and use of public water surface, the collection of the fees over occupation and use of public water surface, the prior approval of the action plan and the compensation scheme.

The new and renewed construction of the offshore wind power facilities needs the permit for the occupation and use of public water surface for up to thirty years, conditions of which can be added to the permit by the conditions for the protection of the marine environment, ecosystems, fishery resources and natural landscape and prevention of fishery damage.[40]

The authorities managing the public water surface can collect the fees over occupation and use of public water surface from the energy production business operators.[41] However, the fee can be reduced up to 50% over occupation and use of public water surface for establishment and operation of the offshore wind power facilities.

Besides the fee, the permitted entrepreneur shall compensate over the loss caused by reclamation of public water surface.[42] Also, the licensee for reclamation shall

---

35. *The Marine Environment Management Act*, Act No. 13383, June 22, 2015, Art. 57.
36. *Id.*, Art. 84.
37. *Id.*, Art. 85.
38. *Id.*, Art. 79.
39. *The Public Waters Management and Reclamation Act*, Act No. 13385, June 22, 2015.
40. *Id.*, Arts 8, 10.
41. *Id.*, Art. 13.
42. *Id.*, Art. 31.

compensate fisheries loss or provide facilities to prevent damage.[43] That compensation should be done prior to construction by market value.

[5] Habitat and Species Protection

The Act on the Conservation and Management of marine ecosystems[44] regulates the designation and management of the protected marine area, limiting the activities in the protected marine area and the collection of the cooperation charge for marine ecosystems conservation. Within the protected marine area, construction and extension of the new buildings and artificial structures is limited. Also, the Minister of Oceans and Fisheries collect a cooperation charge for marine ecosystems conservation over the development projects subjected to the EIA that are operated within the public water surface such as offshore wind farm projects.[45]

[6] Fisheries

The Fisheries Act[46] regulates the compensation scheme for a fisheries loss. The person who experiences damage can submit a compensation claim on the fisheries damage with the proof of the fisheries loss within three months from the date of permit or other administrative actions.[47] The compensation should be done before starting construction. The parties who shall compensate are the authorities which give the permit relating to the limitation of fisheries and the beneficiaries from such permits or administrative actions. The authorities decide the amount of compensations and payments through the review of the committee of fisheries within sixty days since the date of claim.[48]

[7] Public Participation Scheme

At least, regarding interest sharing from the wind power development or participation in shares, there are no official programs systemized by law in Korea. Instead, there are community support programs and subsidies to local areas affected by operations of wind power plants.

In Jeju, in 2012, there was a controversial issue when the Jeju-do asked 17.5% of the share interests of the wind power generation companies for the regional development fund when giving permits.

According to the Jeju-do Ordinance on the Wind power business license, the wind energy resource is public common goods of Jeju island and the governor of Jeju

43. *Id.*, Arts 31, 32.
44. *The Act on the Conservation and Management of marine ecosystems*, Act No. 13383, June 22, 2015.
45. *Id.*, Art. 49.
46. *The Fisheries Act*, Act No. 13385, June 22, 2015.
47. *Id.*, Art. 81.
48. *The Presidential Decree on the Fisheries Act*, Decree No. 26707, December 10, 2015, Art. 67.

shall make an effort to let the interests earned by the wind power development projects shared with all the citizens of the islands.[49] Yet, the central government did not accept this kind of contract between the company and the local government.

[8] Grid

In Korea, even though *The Act for Promotion of Smart Grid Deployment and Use*[50] regulates on development and operation of smart grid systems, it does not give any affirmative treatment for offshore wind powers.[51] However, for energy efficiency, the smart grid system provides better infrastructure for the new and renewable energy development as well.

For small scale renewables, it was a hurdle to be connected to the grid. Yet, recently the government allowed unlimited connection to the grid even for the small scale renewables under 1 MW.

Besides direct grid connection, according to the recent new energy investment plan in 2016, the Korean government suggested diverse incentives including the special price for ESS energy supply to promote ESS, which stabilizes variable power generation regardless of weather and wind. Especially, the REC multiplier for ESS facilities connected with wind power facilities is 5.0, which is much higher than that of other facilities.

§9.04 CHALLENGES AND SOLUTIONS

[A] Institutional Design: Challenges and Solutions

One of the hurdles in promotion of the offshore wind energy in Korea is ambiguity of which institute is charge of the management of public water surface. This problem makes it difficult for the offshore wind power projects to get permits. For the use of coastal area inside the exclusive economic zones, it is required to get a permit from the local government. Yet, the boundaries among the areas where the local governments govern are not clear and subject to the customary boundaries. The Korean government had often failed to set up clear lines on water management areas since the controversies among regions and fishermen lead to severe protests.

[B] Incentives: Challenges and Solutions

One of the biggest challenges in developing the offshore wind power in Korean is to overcome regional complaints. Most complaints come from concerns over damage to

49. *The Local Ordinance of Jeju on Wind Power Production Permit and Site Designation Business*, Local ordinance No. 802-3, October 12, 2011.
50. *The Act for Promotion of Smart Grid Deployment and Use*, Act No. 12154, January 1, 2014.
51. As a result, the developers shall pay for the cost for grid connection. The Wind Power Industry Association had proposed that government should support the grid connection or the expense shall be shared through the electric price.

the fisheries. In fact, reasons why social acceptance is very low can be interpreted in many ways; it can depend on the previous experience of regions, misunderstanding on the impact of offshore wind farms, distrust of the government or the public electric company.

Especially, one reason behind low acceptance can be attributed to the fact that there are no regional interest sharing programs other than community support programs provided through the Fund of Electric Industries' Foundation. Thus, it is required to develop participation programs to increase social acceptance besides compensation scheme. When it comes to the participation programs, besides sharing interests they can include providing monetary incentives to local governments or community to do the environmental and marine impact assessments themselves.

Another profound obstacle is the government's unstable policy. It is constantly needed to receive policy support in growing industries like renewables until the market begins to automatically generate benefits. Yet, the policy stableness is not enough to boost offshore wind power industries and develop the technologies. This leads to enterprises becoming reluctant to participate and invest.

Furthermore, the high cost for grid connection from offshore wind turbines is an additional burden laid on developers. Considering the developing level of industries and the dispersed characteristics of the power generators, it is required to make the grid connection less burdened by government support.

[C] Regulations: Challenges and Solutions

In Korean legislations and regulations, the special features of offshore wind energy development are not fully considered although REC gives higher value for offshore wind power. For example, the cost of grid connections is much higher than other energy resources: the neighboring communities impacted are living physically far compared to other energy developments on land and the features of impact are different from those on land.

Even though there is the legally required environment impact assessment, since it is conducted by the constructors, it is needed to assess the environmental and marine impact more objectively and sufficiently in time. Also, there are problems in assessing the damages by fisheries, which often fail to be proved. So, a more systemized database is needed to evaluate the value of coastal and marine use as well as the damages from their loss.

CHAPTER 10

Legal Framework to Develop Offshore Wind Power in Taiwan

Anton Ming-Zhi Gao[*]

§10.01 INTRODUCTION

In order to deal with climate change, Taiwan passed the Green House Gas (GHG) Reduction and Management Act (GHG Act)[1] in mid-2015. One of the key features of this Act is to adopt an EU-like emission-trading scheme (ETS) in Taiwan to price carbon and GHG.[2] In addition, an ambitious emission reduction target is also embedded in this Act. Article 4 of the Act provides that: "Long-term national GHG emission reduction goal shall be to reduce GHG emissions to no more than 50% of 2005 GHG emission by 2050." Such a long-term target would further split into a five-year phase target and sectoral target, respectively.[3] In order to achieve this emission reduction target, the GHG Act also requires the government to develop a National Climate Change Action Guideline and GHG Reduction Action Plan.[4]

In fact, long before the passing of the GHG Act, Taiwan adopted a comprehensive policy package to promote low carbon energy, comprising the 2008 Sustainable Energy

* This article was funded by the National Energy Research Program Phase II, Ministry of Science and Technology, Research Project: Removing the Legal Barriers to the Deployment of Low-Carbon Energy Technologies (2015) Project Number: (104-3113-F-007-001).

A portion of this chapter was previously published in 2015 as "Europe's Policy Framework for Promoting Offshore Wind Energy: Lessons for Taiwan and Other Countries" in Renewable Energy Law and Policy Review, 6(3), by Anton Ming-Zhi Gao. This chapter has been updated to reflect the developments in Taiwan as of July 2016.

1. Greenhouse Gas Reduction and Management Act, 2015.07.01: http://law.moj.gov.tw/Eng/LawClass/LawContent.aspx?PCODE=O0020098.
2. Articles 18, 20, and 21 of the GHG Reduction and Management Act of 2015.
3. *See* Art. 9 of the GHG Reduction and Management Act of 2015.
4. Article 9(1) of the GHG Act.

Policy Framework[5] and the 2010 Master Program on Energy Conservation and Emission Reduction.[6] After the Fukushima accident, a new energy policy was announced by the president at the end of 2011,[7] and a more aggressive approach and detailed programs were proposed and adopted, i.e., Million Solar Roof and the *1,000 Onshore and Offshore Wind Power* (OWF) *Turbines Program*.[8] From the development trend, we can see that the main focus of Taiwan in developing renewable energy sources (RES) is on photovoltaic (PV) and wind power. If one looks closer, there seems to be a trend from onshore to offshore. The rationale for such a change could primarily be related to the already large development of onshore wind, which leads to limited land space upon which to put the wind turbines. So far (August 5, 2016), there have been 336 onshore wind turbines (666.2 MW).[9] In addition, a serious protest regarding the safety of the onshore wind turbine project worsens the development position of onshore wind further.[10] With the renewable electricity goals for 2030 increasing from 12,502 MW (in 2011), to 13,750 MW (in 2014), and then to 17,250 MW (in the recent deliberation of the Energy Development Policy [draft] in 2015),[11] wind power, and particularly offshore wind, is expected to play a major role. In addition, offshore wind has recently been seen by the government as a new energy business for future Taiwan. All of these lead to the move from onshore to offshore.

In the original 2011 version of the 1,000 Onshore and Offshore Wind Turbines Program, the schedule is to complete four offshore wind demonstration turbines by 2016, 1,200 MW onshore and 520 MW offshore wind power plants by 2020, and 3,000 MW between 2020 and 2025, with a total capacity of (around 1,000 wind turbines) 4,200 MW.[12] Such a policy design is a much-enhanced version of the First phase of the Offshore Wind Demonstration Program of 2007.[13] In this program, only the quantitative direction, such as "the efforts to facilitate the deployment of offshore wind," the potential licensing principles, the preliminary review of current Taiwan's R&D situation, etc., are provided. Such an R&D-oriented approach was repeated by the 2007 Energy Technology R&D White Paper[14] and the 2012 Energy Industry Technology

5. MOEABOE, Sustainable Energy Policy Framework 2008 永續能源政策綱領: http://web3.moeaboe.gov.tw/ecw/populace/content/wHandMenuFile.ashx?menu_id = 2154.
6. 2010 Master Program on Energy Conservation and Emission Reduction (行政院節能減碳推動會國家節能減碳總計畫 （核定本）行政院節能減碳推動會秘書處 （經濟部） 中華民國 99 年 5 月): http://www.aec.gov.tw/webpage/other/files/index_04_1_7.pdf.
7. 2011 New Energy Policy, http://www.ey.gov.tw/policy4/cp.aspx?n = E4707ED5C6C73F4B.
8. https://www.moea.gov.tw/Mns/populace/news/News.aspx?kind = 1&menu_id = 40&news_id = 24948.
9. http://www.twtpo.org.tw/.
10. http://www.chinapost.com.tw/taiwan/local/miaoli/2013/05/03/377592/Lawmaker-leads.htm.
11. Energy Development Policy (能源開發政策).
12. http://www.twtpo.org.tw/intro.aspx?id = 462; http://www.twtpo.org.tw/intro.aspx?id = 9.
13. 第一階段設置離岸式風力發電廠方案, http://www.rootlaw.com.tw/LawContent.aspx?LawID = A040100151004000-0960824.
14. MOEABOE, 2007. 2007 Energy Technology R&D White Paper (in Chinese). Available at: http://web3.moeaboe.gov.tw/ECW/populace/content/ContentLink.aspx?menu_id = 473 (accessed on August 15, 2016).

White Paper.[15] Yet, the legal and policy development between 2011 and 2016 evolved more than that between 2006 and 2011.

The purpose of this chapter is to investigate the legal framework of Taiwan in facilitating the deployment of OWF. For this purpose, this chapter will first provide an overview of the institutional design. Then, this chapter will identify the comprehensive legal elements, including incentive and regulations. Finally, this chapter will analyze their weakness and figure a way out and a resolution.

§10.02 INSTITUTIONAL DESIGN FOR OFFSHORE WIND POWER

Institutional design can play the role of the driving engine for further development of policy, incentives, and regulations.

[A] The "Main" Authority During the Development of Offshore Wind Power

The key authority in drafting and preparing policy and law for OWF is mainly the Bureau of Energy, Ministry of Economic Affairs (MOEABOE). The MOEA is an economics ministry and mainly in charge of the function of economic development and industrial development. As the energy issue is of great import regarding economic affairs, the BOE was established by the Act of the Organization of the Bureau of Energy in 2004.[16] The BOE is mainly in charge of energy policy and legislation. As for the OWF-related legislation, the most important one is the *Renewable Energy Act of 2009* (RE Act) and its related ordinances. The second most important legislation is the (non-electricity-liberalization) *Electricity Business Act*, since the OWF developers are subject to the electricity-generation business license requirement.[17]

With regard to technology R&D, Taiwan is a latecomer, having only recently adopted wind farm technology (not to mention the more advanced OWF technology). The government, and particularly the BOE, has conducted considerable research on these technologies in the past decade. For example, a five-year project on the "demonstration and promotion of wind power [from] 2000–2004"[18] was conducted by the Industrial Technology Research Institute in Taiwan (ITRI), while a special research project on offshore wind farms, "The Development of Offshore Wind Power Technology," was conducted from 2009–2010.[19]

15. MOEABOE, 2012. 2012 Energy Technology R&D White Paper. Available at: http://web3.moeaboe.gov.tw/ECW/populace/content/SubMenu.aspx?menu_id=62 (accessed on August 15, 2016).
16. http://web3.moeaboe.gov.tw/ECW/english/content/Content.aspx?menu_id=963.
17. http://law.moj.gov.tw/LawClass/LawContent.aspx?PCODE=J0030012.
18. Wind Power Demonstration Project (2000–2004), ITRI, research projects issued by the Energy Commission and (later on) the Bureau of Energy, Ministry of Economic Affairs.
19. Technology Development of Offshore Wind Power Generation (離岸式風力發電技術開發), 2009–2010, ITRI, research projects issues by the Bureau of Energy, Ministry of Economic Affairs. *See* Anton Ming-Zhi Gao, "Europe's Policy Framework for Promoting Offshore Wind Energy:

Even if the MOEA and the BOE's main focus is on economic and energy affairs, the attention given to climate issues is gaining ground under this ministry as well. On June 16, 2006, the Taiwan Industrial Greenhouse Office (TIGO) was established by the Taiwan MOEA in an effort to facilitate industrial greenhouse gas management and promote voluntary reduction as well as to ensure the competitiveness of our industries.[20] This office has been renamed the "Greenhouse Office of MOEA (GO-MOEA)" in 2008 and "Energy-Saving and Carbon Emission Reduction Office, Ministry of Economic Affairs" in 2010, respectively. The promotion of OWF is considered to be a converging policy area for economic growth, new industry, and achieving climate-change goals. Aside from the missionary office created to address climate change issues, there are several *Project-Based Missionary Offices* related to the development of OWF:

- Renewable Energy Feed in Tariff (FIT) Office, run by the Taiwan Institute for Economic Research (TIER).
- 1,000 Wind Power Projects Office, run by the ITRI.
- Renewable Energy Promotion 4 into 1 Office, run by the ITRI.

[B] The "Supplementary" Authorities During the Development of Offshore Wind Power

[1] Central Government Level

Apart from the MOEA and the MOEABOE, there are also other institutions related to the development of renewable energy policy. At the cabinet level, climate and energy-related missionary commissions have been established since 2009, including: the Energy Conservation and Emission Reduction Commission, the Executive Yuan (since 2009); the New Energy Promotion Commission, the Executive Yuan (since 2009); the Green Energy and Low-Carbon Promotion Commission (since 2014); and the Office for Energy Conservation and Emission Reduction, (since 2016). These types of institutions have been in charge of the development of a cross-ministerial policy package, i.e., the Master Program on Energy Conservation and Emission Reduction, and of coordinating the efforts and functions of different ministries in related climate issues, such as drafting Intended Nationally Determined Contributions (INDC) and deliberation on Nationally Appropriate Mitigation Actions (NAMAs). Recently, the newly established Office for Energy Conservation and Emission Reduction has been working very actively to coordinate the fragmented legislation of different ministries.

The Ministry of Science and Technology (MOST) is also playing a key role in OWF planning. Additionally, the Master Program on Offshore Wind Power, under the National Science and Technology Program-Energy (NSTPE), will promote offshore wind technology research from 2013–2015.

Lessons for Taiwan and Other Countries" in Renewable Energy Law and Policy Review, 6(3), at 12.

20. https://www.go-moea.tw/en/.

At the legislation level, the authorities listed in Table 10.1 are relevant during the development stage of OWF, which was identified by the 2007 OWF Program,[21] but remains unresolved.

Table 10.1 Authorities Relevant to OWF Development

Matter	*Authority*
Establishment of electricity business	Ministry of Economic Affairs (Bureau of Energy)
Laying of undersea electricity cable	Ministry of the Interior (Department of Land Administration)
Artificial islands and structures	Ministry of the Interior (Construction and Planning Agency)
State-owned and non-public land	Ministry of Finance (National Property Administration)
National security-related regulations and construction-prohibited matters	Coast Guard Administration, Executive Yuan (Ministry of Defense)
Flight safety	Ministry of Transportation
Shipment safety	Ministry of Transportation
Environmental impact assessment	Environmental Protection Administration
Marine pollution	Environmental Protection Administration
Mining rights	Ministry of Economic Affairs (Bureau of Mining)
Fishing rights	Agricultural Commission, Executive Yuan (Fishery Agency)

Source: MOEABOE, 2007a. The program for the first stage of offshore wind development, approved by the Executive Yuan in August 2007 (in Chinese). Available at: http://web3.moeaboe.gov.tw/ECW/populace/content/Content.aspx?menu_id=1077 (Accessed on November 15, 2015).

[2] The Role of the Local Government

The role of the local government is limited under the RE Act of 2009. This situation applies in OWF development, as well. The main policy, incentives, and regulations for OWF are mainly dealt with at the central government level. Additionally, the grid planning[22] and harbor[23] are also among the central government's duties.

21. MOEABOE, 2007a. The Program for the First Stage of Offshore Wind Development, approved by the Executive Yuan in August 2007 (in Chinese). Available at: http://web3.moeaboe.gov.tw/ECW/populace/content/Content.aspx?menu_id=1077 (accessed on August 15, 2016).
22. The grid is mainly planned by Tai-Power, a vertical integrated electricity company.
23. Ordinance on the Planning, Construction, and Operating of Industrial-Exclusive Harbor (工業專用港或工業專用碼頭規劃興建經營管理辦法) http://law.moj.gov.tw/LawClass/LawContentIf.aspx?PCODE=J0030104.

The local government's role is mainly provided in the Electricity Business Act, where the application for setting up an electricity-generation business (more than 500 kW) requires the consent of the local government under the Electricity Business Licensing Ordinance.[24] In addition, the environmental regulation regarding grid development may be relevant to the OWF. Yet, the role of the local government in OWF development has been increasingly gaining importance recently as the new legislation is adopted. For instance, the Coastal Zone Management Act of 2015[25] and the Spatial Planning Act of 2016[26] confer authority to the local government. Under the GHG Act, the local government also has the power to propose a local plan for emission reduction[27] where the OWF could play a potential role.

[3] Other Players Vital for the Development of Offshore Wind

During the OWF R&D process, many energy thinking tanks, such as ITRI, TIER, and the Taiwan Research Institute (TRI), are involved in providing advisory opinions to the BOE and related government agencies. In addition, the state-owned electricity company, Tai-Power, also has a duty to help launch the OWF demonstration.

In order to provide a platform to encourage discourse among industry, government, academia, and research stakeholders, the Taiwan Wind Energy Association has been very actively promoting OWF, as well.[28]

Finally, the Taiwan OWF developers are more active in OWF than in the onshore wind area. As for the onshore wind developers in Taiwan, there are only two: one is a private company, Infravest, importing Germany's Enercon system, and the other is dominated by state-owned Tai-Power. Yet, the OWF policy draws a great deal of attention from Taiwan private developers, such as Taiwan Generations Corp.[29] and Swancor.[30]

[C] The Role of "Law" in Institutional Decision Making and Related Policies

Since the first announcement of OWF-related research projects in 2000, the main focus in the early stage (2000–2007) has been the OWF policy and the development of the Renewable Energy Bill (since 2001). The aspect of law is not focused in OWF but rather the general incentive framework for all renewables.

After the adoption of the RE Act in 2009, the focus began to shift to its economic aspect, i.e., the rate of the feed-in tariff. On the first 2010 rate schedule to implement the RE Act, the OWF rate was set at NTD 4.1982/per kWh (around 12 cents if using the

24. Article 3 of the Electricity Business Licensing Ordinance. http://law.moj.gov.tw/LawClass/LawAll.aspx?PCode=J0030012.
25. http://law.moj.gov.tw/Eng/LawClass/LawContent.aspx?PCODE=D0070222.
26. http://law.moj.gov.tw/Eng/LawClass/LawContent.aspx?PCODE=D0070230.
27. Article 15 of the GHG Reduction and Management Act.
28. http://www.twnwea.org.tw/.
29. Taiwan Generations Corp. www.taiwangenerations.com/.
30. http://www.swancor.com/tw/index.php.

currency exchange rate of 1:35);[31] the yearly rate-setting events have been the main battleground for all renewables, including OWF. As the rate is too low to promote OWF, it has been enhanced from 4.1982 in 2010 to 5.7405 in 2016. An alternative front-end rate is also possible (the first 10 years at 7.1085, the other 10 years at 3.4586).[32]

The National Energy Technology Research Program (NEP) hosted by MOST (and its former National Science Council) should not be ignored. During the first phase of the research (2009–2013), the comprehensive legal framework for specific renewable electricity was a research topic. It is helpful to move the OWF law beyond just rate design, per se. The latter development of the Strategic Environmental Assessment (SEA) of OWF in 2015[33] and the OWF Zonal Development Ordinance of 2015 (Ordinance on the OWF Planning Sites)[34] could be affected by the results of the first phase legal research conducted by our team.[35]

The OWF legislation concerns began to be integrated into the decision-making process after the 2011 new energy policy proposed following the 2011 Fukushima accident. Under this policy, the aforementioned 1,000 Onshore and Wind Power Turbine Program was proposed, and a missionary projected based Program Office was established. Before this time, even though the First Phase OWF Program already highlighted the bureaucracy issue in 2007, laws tackling the issue have experienced limited progress. The same also applies to the non-FIT and regulatory barriers of the OWF. Thus, the law begins to be integrated more into the decision-making process. The 2012 OWF Demonstration Subsidy Ordinance[36] could be seen as the early achievement of such efforts.

In the second phase of the National Energy Technology Research Program, the role of law is largely enhanced. A special office, i.e., the Office of Energy Policy for Bridging and Communication, was created under the National Energy Research Program to attempt to deal with the fundamental issues, such as inter- and intra-government communication, legislation, and public communication/participation issues. In 2016, the Director of the National Energy Technology Research Program, Dr. Lee, was promoted as the Minister of the MOEA, and the new director, Dr. Yang, was appointed as the head of the Office for Energy Conservation and Emission Reduction. There is likely to be a platform for the legislation to be integrated into the decision-making process. This is also a good example of a positive integration of the two.

31. http://web3.moeaboe.gov.tw/ecw/populace/Law/Content.aspx?menu_id = 1083.
32. http://web3.moeaboe.gov.tw/ecw/populace/Law/Content.aspx?menu_id = 2977.
33. EPA Approves SEA on Offshore Wind Energy Development, Taipei Times, www.taipeitimes.com/News/taiwan/archives/2016/07/14/2003651002.
34. Ordinance on the OWF Planning Sites 離岸風力發電規劃場址申請作業要點.
35. 落實我國「國家適當減緩行動」所需法規建制或調適之研究 (The Establishment of Legal Framework to implementing NAMAs [Nationally Appropriate Mitigation Actions]), 2011, 2012, 2013.
36. OWF Demonstration Subsidy Ordinance (風發電岸系統示範獎辦法) http://web3.moeaboe.gov.tw/ecw/populace/Law/Content.aspx?menu_id = 1850.

§10.03 LEGAL DESIGN FOR OFFSHORE WIND

[A] Incentives

[1] Main Finance Scheme

Since the 2009 RE Act, to promote overall renewable energy development, Taiwan passed the RE Act of 2009, which is similar to the German RE Act of 2000. This act mainly promotes OWF development via FIT, with a rate of NTD 4.1982/kWh (for twenty years) in 2010,[37] later increased to NTD 5.5626/kWh,[38] and it increased to NTD 5.7011/kWh in 2016.[39] Starting in 2016, OWF developers will be allowed to choose a fixed tariff of 5.7011 for twenty years or a front-end tariff (7.0035 for the first ten years, 3.4446 for the next ten years). The latter option is also good for the finances of OWF developers.

[2] Supplementary Finance Scheme

[a] Low-Interest Loan and Loan Guarantee

Despite the existence of the FIT scheme for several years, it is still not attractive enough to realize the first OWF project. Thus, "The Ordinance on Demonstration Subsidies for Offshore Wind Farms" was adopted in mid-2012.[40] According to Article 5 of the Ordinance, it provides the subsidy for demonstration plants and demonstration wind farms. The limit for the former is mainly 50% of the total cost of such a plant, and it is NTD 250 million for the latter.

This appears to be an investment subsidy or demonstration subsidy, so why does this article not mention such a finance scheme? According to Article 6 of the ordinance, there is a need to return the subsidy gradually from part of the FIT after the commercial operation of the concerned OWF commenced. Thus, this scheme can be seen as an "interest-free loan" instead of a free investment subsidy.

---

37. MOEABOE, 2010. The Ordinance on the Rate Schedule for Renewable Electricity in 2010 (in Chinese). Available at: http://web3.moeaboe.gov.tw/ECW/populace/content/Content.aspx?menu_id=1083 (accessed on August 15, 2016).
38. MOEABOE, 2011. The Ordinance on the Rate Schedule for Renewable Electricity in 2011. Available at: http://www.esdtaiwan.edu.tw/upload/%7BEE832687-2AD7-4D05-B91D-83CC2D3508F8%7D/%E4%B8%AD%E8%8F%AF%E6%B0%91%E5%9C%8B100%E5%B9%B4%E5%BA%A6%E5%86%8D%E7%94%9F%E8%83%BD%E6%BA%90%E9%9B%BB%E8%83%BD%E8%BA%89%E8%B3%BC%E8%B2%BB%E7%8E%87%E5%8F%8A%E5%85%B6%E8%A8%88%E7%AE%97%E5%85%AC%E5%BC%8F.pdf (accessed on August 15, 2016); MOEABOE, 2012c. The Ordinance on the Rate Schedule for Renewable Electricity in 2012. Available at: http://web3.moeaboe.gov.tw/ECW/populace/content/Content.aspx?menu_id=1103 (accessed on August 15, 2016); MOEABOE, 2013b. The Ordinance on the Rate Schedule for Renewable Electricity in 2013. Available at: http://web3.moeaboe.gov.tw/ECW/populace/content/Content.aspx?menu_id=1955 (accessed on August 15, 2016).
39. http://www.chinatimes.com/newspapers/20151010000037-260202.
40. http://web3.moeaboe.gov.tw/ecw/populace/Law/Content.aspx?menu_id=1850.

Such a subsidy, however, may not be sufficient. Therefore, certain private offshore wind developers complained about the lack of loan schemes, even at the end of 2015.[41] Thus, extra loan resources remain necessary. This year, on May 10, the private company, Swancor, received a loan of NTD 2.5 billion from a bank consortium of Cathay Bank, En-Tie bank, and BNP Paribas.[42] This is also the first case in Taiwan to follow the Equator Principles.[43]

[b] Tax Credit

Article 16 of the RE Act is also likely to promote tax incentives, such as the exemption of the OWF excise duty.[44]

[c] Investment Subsidy/Grant

As FIT is in place, there is no investment subsidy provided.

[d] Favorable Grid Cost-Sharing Rules

This element is very important for the cost of the OWF project. According to estimates, grid cost could account for 25% of the entire project cost of OWF.[45] Thus, if this cost is not taken into consideration, the subsidy focusing merely on offshore wind turbines themselves may be underestimated and insufficient.

Among all connection rules, the super-shallow connections scheme is widely accepted as the best regime for promoting OWF.[46] Under such an arrangement, all costs are socialized via tariff, with no costs charged to the connecting entity.[47]

---

41. http://www.cw.com.tw/article/article.action?id = 5073372.
42. https://tw.mobi.yahoo.com/home/%E5%8F%B0%E7%81%A3%E9%A6%96%E4%BB%B6%E8%B5%A4%E9%81%93%E5%8E%9F%E5%89%87-%E9%A6%96%E5%AE%97%E9%9B%A2%E5%B2%B8%E9%A2%A8%E9%9B%BB%E8%9E%8D%E8%B3%87%E6%A1%88-%E5%9C%8B%E6%B3%B0%E4%B8%96%E8%8F%AF%E9%8A%80%E4%B8%BB%E8%BE%A6-105845136.html. For the website of Cathay Bank, please *see*: https://www.cathaybk.com.tw/.
43. Equator Principles, www.equator-principles.com/.
44. Article 16(1) of the RE Act: "Where construction or operation machineries, equipment, special means of transport for construction use, training materials and such required components imported by corporate legal persons for the construction or operation of renewable energy power facilities are proved by the central competent authority that such use is verified and not domestically manufactured or supplied, import tariffs shall be exempted."
45. Sascha T. Schrödera and Lena Kitzinga, 2012. The Effects of Meshed Offshore Grids on Offshore Wind Investment: A Real Options Analysis. Available at: http://proceedings.ewea.org/annual2012/allfiles2/1356_EWEA2012presentation.pdf: p. 1 (accessed on August 15, 2016).
46. Hans Auer, 2006. The Relevance of Unbundling for Large-Scale RES-E Grid Integration in Europe, Energy & Environment, 17(6), 907, 928; Rüdiger Barth et al., 2008. Distribution of Costs Induced by the Integration of RES-E Power, submitted to Energy Policy, Manuscript No. JEPO-D-07-00521.
47. ENTSO, 2013. ENTSO-E Overview of Transmission Tariffs in Europe: Synthesis 2013. Available at: https://www.entsoe.eu/publications/market-reports/Documents/Synthesis_2013_FINAL_04072013.pdf, p. 35 (accessed on August 15, 2016). Environmental Protection Administration

However, such a preferential model is not adopted in the RE Act of 2009. According to Article 8 of the act, even though there is a mandatory grid connection duty for the grid company,[48] the connecting entity may have to pay for the cost of connection[49] and expansion.[50] There is not a different cost-sharing rule regarding grid connection and expansion for OWF. Thus, it would be considered a "shallow connection or deep connection" and unfavorable for the OWF.

[e] The Links to ETS, CDM, and JI

The GHG Act was adopted in 2015. In this Act, the ETS was planned. Thus, the emission reduction value of OWF would be relevant. Additionally, the OWF development may get extra funding from the GHG Fund. Article 19 of the GHG Act plans to set up a GHG Management Fund, and the funding raised partly by the tendering may be used to facilitate the development of the OWF.[51] This fund and function is very similar to the EU NER 300.[52]

(EPA), Taiwan, 2006. Ordinance on the List of Must-SEA Policy, Plan, and Program (in Chinese). Available at: http://law.epa.gov.tw/zh-tw/laws/868442060.html (accessed on August 15, 2016).

48. Article 8(1) of the RE Act: "The renewable energy power facilities and the electric power generated shall have the stability of their power grids evaluated by local electric power grid enterprises and have them paralleled and bought wholesale at the locations where existing power grids are closest to renewable energy power assembly sites and provide electricity required by such power facilities during maintenance shutdown period; electricity enterprises shall not reject the aforesaid request without proper reason and approval of the central competent authority."
49. Article 8(4) of the RE Act: "The installer shall install and maintain the circuits connecting renewable energy power facilities and power grids; the electricity enterprises with which its power facility are paralleled to the installer's facilities shall provide assistance when necessary; the renewable energy power generation facility installers shall bear the costs incurred."
50. Article 8(2) of the RE Act: "...other than the existing circuits, the cost for power grid enhancement will be shared among both the electricity enterprises and renewable energy power facility installers....".
51. Article 19 of the GHG Act: "the central competent authority shall establish the GHG Management Fund (Fund) from the following sources:

    1. Proceeds from allowances auctioned or sold pursuant to the foregoing Article;
    2. Fees collected pursuant to Article 21;
    3. Government grant via budget appropriation;
    4. Revenues collected under fines and penalties prescribed in the Act;
    5. Money received from persons, liable entities or organizations; and
    6. Other incomes.

    The Fund shall serve the following purpose only for GHG emissions reductions and adaptation to climate change:

    1. Reduce GHG emissions;
    2. Inspect emission sources;
    3. Provide emission sources with assistance, subsidies and grants for voluntary efforts to reduce GHG emissions;
    4. Administrative affairs of holding accounts establishment in the Registry, auctions, sales and allowance trading;
    5. Employ staff to carry out administrative services;

[f] *Other Funding Sources, Such as the National Research Fund and Supportive Scheme*

Taiwan has undergone a long stage of OWF R&D. During this period, the main funding source has been the government's R&D budget. More specifically, a large amount of funding came from the Energy R&D Fund (much like public benefit funds) under the Energy Management Act.[53] At this stage, due to the lack of funding, demonstration projects are unlikely. This also led to the aforementioned adoption of FIT in 2009.

[g] *International Trade Law Concerns Regarding the Subsidy*

One of the main purposes of OWF policy in Taiwan is to cultivate a new OWF business for the nation. In order to improve the competitiveness of Taiwan's industry during the subsidy process, there has been discussion about local content requirements or local manufacturing promotion measures.[54] Yet, there are also concerns about the compatibility of such a local protection regime with the General Agreement on Tariffs and Trade (GATT). The violations of Article 2.1 of the Agreement on Trade-Related Investment Measures (TRIMs), of Article 3.4 of GATT 1994 in the Canadian case of the Green Energy Green Economy Act, Ontario, and in the Indian case of Jawaharlal Nehru National Solar Mission, have been heavily cited and discussed in Taiwan.[55] In order to strike a balance between local development requirements and international trade law, Taiwan adopts a relatively lenient approach to protecting its local industry. Unlike the use of local content requirements as the criteria to be eligible for the main finance

---

6. Coordinate, plan and implement adaptation to climate change;
7. Educate, promote, and award grants;
8. Conduct international affairs;
9. Carry out research and analysis."

52. http://ec.europa.eu/clima/policies/lowcarbon/ner300/index_en.htm.
53. Article 5 of the Energy Management Act: "The central competent authority may establish a special fund for research and development of energy together with a project plan for the purpose of enhancing research and development of energy in accordance with the Budget Law.

    The foregoing special fund shall serve the following purposes:

    1. Research and development of technology relevant to exploitation of energy resources and alternative energies.
    2. Research and development relevant to technology and methodology for the rational and efficient use of energy as well as energy conservation.
    3. Economic analysis and collection of information on energy.
    4. Training of experts in energy planning and technology.
    5. Other expenses as approved.

    Incentives or subsidies may be granted to the juristic person or individual whose engagement in research under subsection 1 and 2 of the preceding paragraph proves to be highly practical. The central competent authority shall submit to the Legislative Yuan the annual report on the effectiveness in carrying out the energy research and development plan and of the use of the special fund."
54. *See*, for example, http://www.twtpo.org.tw/offshore_faq.aspx.
55. *See*, for example, A Study Report on the India PV Subsidy Cases (簡析美國針對印度國家太陽能計畫提起諮商案). *www.tradelaw.nccu.edu.tw/epaper/no145/2.pdf.*

scheme, such as FIT or investment subsidy, the regime in Taiwan is relatively lenient. Article 18(2) of the OWF Subsidy Ordinance gives competence to the MOEA to ask the state-owned electricity company to prioritize local manufacturing principle and states that *two* local-made wind turbines may be deployed, while related laws should be respected. In addition, according to Article 18(3) of the OWF Subsidy Ordinance, the construction process of these two wind turbines is allowed to be more flexible than that of the others.

[3] Incentive for Construction of Harbor, Vessel, and Grid, or for Turbine Manufacturers

With a view to finishing the construction of OWF, the harbor and vessels are all "essential" facilities. Yet, in Asia, there are few countries with such technology and facilities, which thus heavily require assistance from Europe or other countries. In addition, even at the stage of the R&D process, such vessels are necessary for the further planning of real construction. Thus, in the past few years, there have been repeated incidents of the controversial use of China's working ships in the Taiwan Strait.[56] This controversy has also led to the biggest Taiwan shipbuilder, CSBC, investing in construction vessels[57] and providing offshore wind service.[58] As CSBC was a state-owned shipbuilding company before its privatization in December 2008, such investment could also have some "state" element behind the scene.

As there is no OWF turbine manufacturer in Taiwan to date, the incentive is only limited to *R&D investment* to attract the interests of local industries. In the past few years, China Steel and TECO decided to work together to be in this market.[59]

Finally, in terms of the harbor, none has been developed yet. However, Chapter 10 of Taiwan's Act on Industrial Innovation provides further incentives for the Establishment and Management of Exclusive Industrial Harbors and Exclusive Industrial Wharfs.[60] The offshore wind harbor could be developed under this Act and its ordinance.[61]

56. Taiwan company withdraws notice to use Chinese vessels over security concerns, http://www.reuters.com/article/us-taiwan-china-energy-idUSKCN0HK0AY20140925; Chinese Ships Back on Offshore Project, Taipei Times, www.taipeitimes.com/News/front/archives/2015/06/10/2003620335.
57. Taiwan Shipbuilder Eyes Offshore Wind Sector, IHS Fairplay, http://fairplay.ihs.com/commerce/article/4268886/taiwan-shipbuilder-eyes-offshore-windfarm.
58. Offshore Wind Turnkey Service Provider, CSBC Corporation, Taiwan, http://www.csbcnet.com.tw/English/ServiceEng/EnergyZoneEng/Energy02Eng.htm.
59. Teco, China Steel Team Up for Taiwanese Offshore Wind, Offshore ..., http://www.offshorewind.biz/2015/07/23/teco-china-steel-team-up-for-taiwanese-offshore-wind/.
60. *See* Arts 56–64 of the Industrial Innovation Act.
61. Ordinance on the Planning, Construction, and Operating of an Industrial Exclusive Harbor (工業專用港或工業專用碼頭規劃興建經營管理辦法.

[B] Regulations

[1] License Scheme

The permit conditions of OWF in Taiwan may not be very clear. The main rules for granting a permit are provided in the Ordinance of Electricity Business Registration (電業登記規則). There have been three stages for granting an electricity-generation license, including the preparatory stage, construction permit stage, and issuing an electricity license stage. At the preparatory stage, "private" developers are required to submit a great deal of paperwork, according to Article 3 of this Ordinance, including:

- Submitting a Preparatory Plan Document
- Applying for an environmental impact assessment (EIA) permit
- Obtaining a letter of consent from the local government
- Obtaining a letter from a bank proving the intention to provide a loan
- Completing a grid connection consent form
- A letter stating an opinion on or consent relating to the relevant flight authorities, radar, military, harbor safety, aquatic life and cultivation protection zones, fishing rights, and mining rights, as well as an application for a permit to explore the possibility of installing an undersea electricity cable.

Apparently, the authorities have wide discretion over these documents. For instance, the empty condition of "obtaining a letter of consent from the local government" may cause difficulties for developers who wish to challenge the decision if a permit is not granted. In addition, the confusing wording of "opinion or consent" may also have polarizing legal effects.

Due to this three-stage procedure and requirement, it usually takes several years to realize an onshore wind project, not to mention a more complex OWF project.

This problem may result from the mixture of electricity-generation licensing and an installation permit. As the Taiwan electricity market is not yet liberalized, there are bothersome restrictions on the establishment of a new generation company that also apply to renewable electricity companies.

[2] Environmental Impact Assessment

The current EIA regime is not engineered for ideal OWF development. For instance, according to Article 29 of the EIA Ordinance,[62] geothermal and marine energy projects are eligible for EIA pilot-test or demonstration exemptions, but OWFs are not, suggesting that EIA could potentially hinder OWF development.[63]

62. EPA, 2012. The Ordinance of Compulsory EIA Projects. Available at: http://law.moj.gov.tw/Eng/LawClass/LawContent.aspx?PCODE=O0090012 (accessed on August 15, 2016).
63. Gao, *supra* note 19, at 13–14.

In addition, the zero threshold for OWFs, similar to that for nuclear power plants, may place too great an environmental burden on the development of OWFs. Apparently, this value in Taiwan does not match that of the EU's EIA Directive. Additionally, as several non-governmental organizations (NGOs) have used EIAs to stall the development of many onshore wind and other renewable energy projects, they might do the same to prevent OWF realization.

Finally, Taiwan's EIA system is very like that of Denmark but not like the ordinary ones under the EIA Directive. Article 14 of the EIA Act provides that: "The industry competent authority may not grant permission for a development activity prior to the completion of an environmental impact statement review or the authorization of an environmental impact assessment report; permission granted in violation of this regulation shall be invalid." Such an EIA veto function[64] included in the Taiwan EIA scheme may place an extra regulatory burden on developers in Taiwan.

[3] SEA and Siting-Related Planning

In Taiwan's original Ordinance in the Mandatory SEA Policy and Program,[65] there is a lack of consideration about combining energy development with spatial planning and land use. Only the Energy Development Policy is subject to mandatory SEA under the energy policy; there is no similar requirement in the land-use policy. Thus, the development of the 1,000 Onshore and Offshore Wind Power Program, as noted above, is not subject to mandatory SEA. This situation is not good news for SEA. As the SEA process usually involves government funding for wind resources analysis and preparatory communication conducted prior to making a project public, it would be very helpful in reducing the financial burden carried by developers at the project level.

After promotion by the legal study under the National Energy Technology Research program, SEA has been implemented. In order to resolve potential site-selection barrier problems, on July 2, 2015, the government promulgated the "Ordinance on the OWF Planning Sites".[66] This ordinance represents an effort to identify high-potential sites and provides for the use of different sites in the case of conflicts. These sites include: fishing harbors, wetlands, set net fishing rights zones, natural gas pipelines, undersea pipelines, electricity grids on Pen-Hu Island, undersea cables, national parks, reef protection zones, coal harbors, power plant zones, shooting ranges, zones where anchoring is prohibited, harbor and anchor zones, important wild bird habitats, important habitats for wild animals, zones where fishing is prohibited (artificial fish reefs, fish resources conservation zones, industrial zones, offshore demonstration sites, important white dolphin habitat zones, and coastal area natural

64. Article 14 of the EIA Act: "The industry competent authority may not grant permission for a development activity prior to the completion of an environmental impact statement review or the authorization of an environmental impact assessment report; permission granted in violation of this regulation shall be invalid."
65. http://ivy5.epa.gov.tw/epalaw/search/LordiDispFull.aspx?ltype=03&lname=1300.
66. http://web3.moeaboe.gov.tw/ECW/populace/Law/Content.aspx?menu_id=2870.

environmental protection zones), and zones where construction of other infrastructure is prohibited.

Even if the nature of such an ordinance is only a weakened version of SEA, a map for each county could provide a measure of certainty for developers. Figure 10.1 shows the example of Hsin-Chu County.

Figure 10.1 Potential OWF Sites in Hsin-Chu County

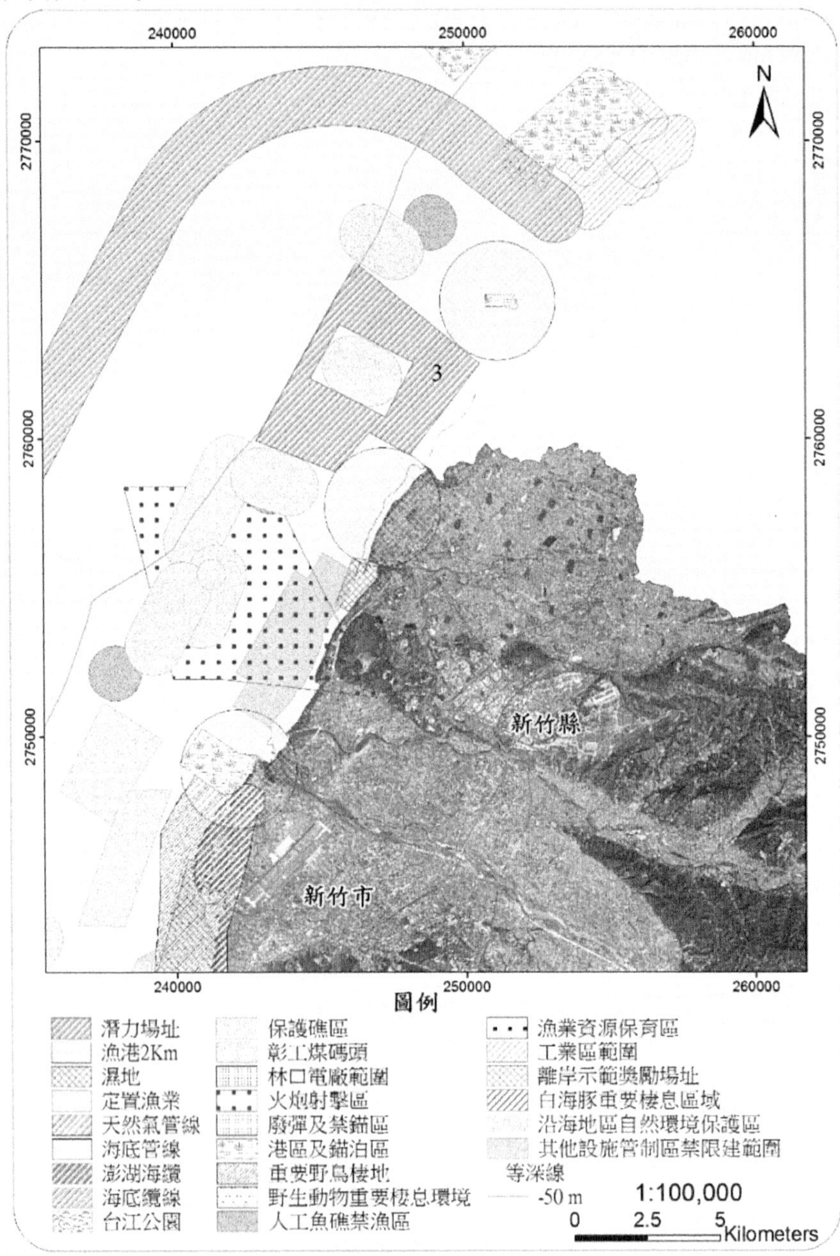

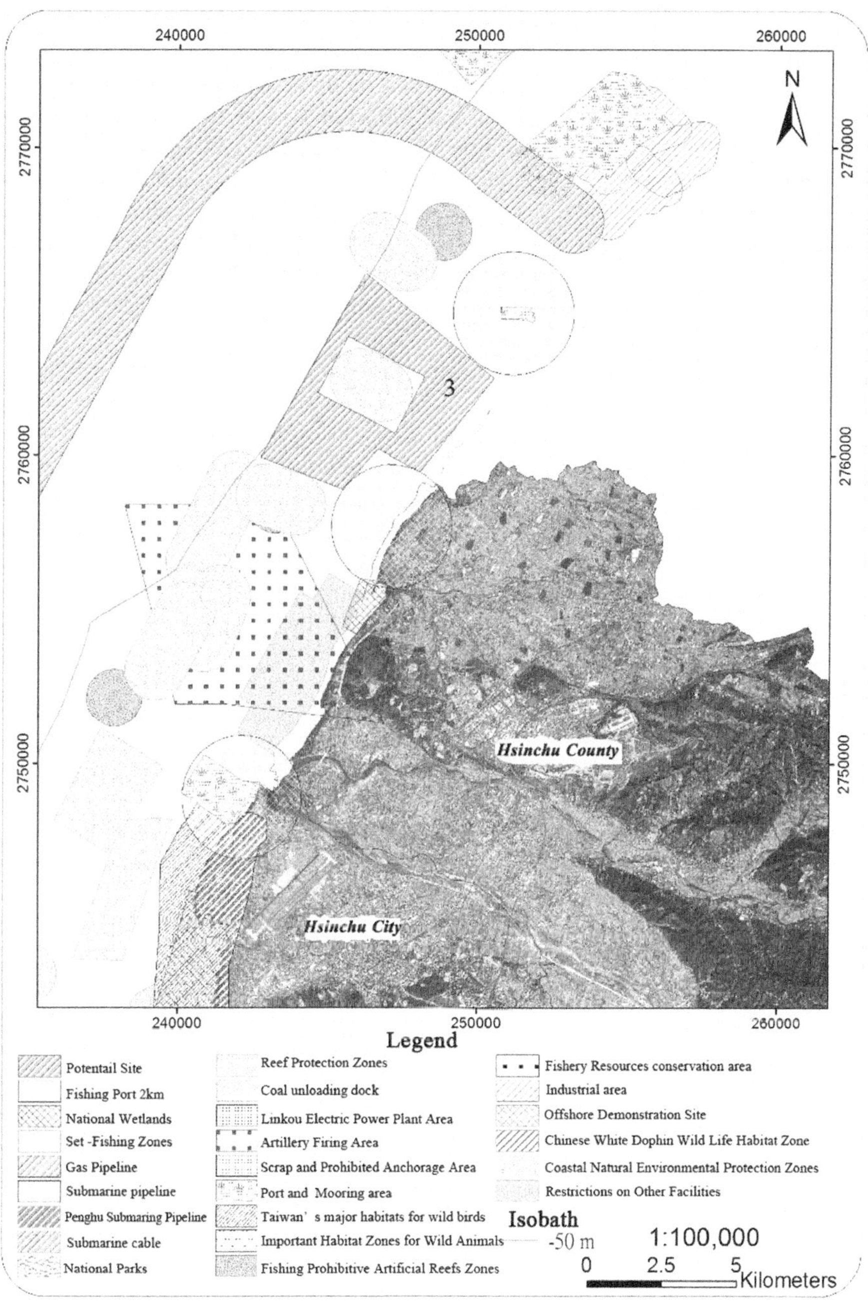

Retrieved from http://web3.moeaboe.gov.tw/ECW/main/Law/wHandEditorFile.ashx?file_id = 1940.

After the publication of the "Ordinance on the OWF Planning Sites," the MOEABOE held a meeting on the policy consultation on October 6, 2015, a scoping meeting on November 18, 2015, proposed a draft SEA report on the ordinance and submitted it to the Environmental Protection Administration (EPA) at the end of 2015.[67] So far, there have been two expert advisory meetings held on March 9 and July 12 of 2016.[68] The SEA report has been conditionally approved by the second meeting and is very likely to be confirmed by the final meeting.[69] The condition provides the following.

The committee decided that offshore wind turbines could not be constructed within 500 m of white dolphin habitats, and construction noise must be kept below 180 decibels; construction must be suspended if white dolphins are discovered near construction sites.

Offshore wind farms cannot be constructed near wetlands or ecologically sensitive areas, either.

The EPA also added the extra condition that the location of OWF should avoid fourteen types of zones, including: national wetlands, set-fishing zones) aquatic animals and plants protection zones, reef protection zones, fishing prohibitive artificial reefs zones, important habitat zones for wild animals, wild animal protection zones, Chinese white dolphin wildlife habitat zones, national parks, coastal natural environmental protection zones, national scenic areas, underwater cultural heritage protected areas, second-degree coastal protection areas, and Taiwan's major habitats for wild birds. Apparently, there would be few locations for OWF after such developing zone exclusions.

[4] Planning the Legal Regime: Marine Planning and/or Land Planning

The main legislation governing the land planning was the Regional Plan Act.[70] Under this act, the Ministry of Interior Affairs (MOIA) and local governments are in charge of preparing the national regional plan and the local regional plan, respectively.[71] Yet, for the *non-urban land*, there are special regulations under "Chapter III: Control of Regional Land Utilization." For instance, after a regional plan is announced and implemented for the non-urban lands, the concerned local governments shall work out non-urban land-use zoning maps.[72] The related cases may be subject to the development under certain situations.[73] Further regulations are provided in the Ordinance on

67. http://law.ndc.gov.tw/Suggestions_detail.aspx?ID = 6015 離岸風電區塊開發政策評估說明書(初稿)開發單位： 經濟部能源局中 華 民 國 1 0 5 年1 月.
68. For information on the two experts' consultation meeting, *see*: http://doc.epa.gov.tw/IFDEWebBBS_EPA/Project/EPA/list03.aspx.
69. EPA Approves SEA on Offshore Wind Energy Development, Taipei Times, July 13, 2016.
70. http://law.moj.gov.tw/Eng/LawClass/LawContent.aspx?PCODE = D0070030.
71. Article 5 of the Regional Plan Act.
72. Article 15 of the Regional Plan Act.
73. Article 15(2) of the Regional Plan Act:

    1. The application is suitable and reasonable for land utilization.

Regulating the Use of Non-Urban Land.[74] The allowed use item and the procedure have been further specified in Article 6(3), 6-2, and the Annex Table of 1-1.

Table 10.2 The Allowed Activities and Regulations of Sea Land under the Ordinance on Regulating the Use of Non-Urban Land

Allowed Activities	*Activities Require Zonal Permission from the Central Competent Authority and the Permission of the Central Industry Authority*
1. Use of fishery resources	...
2. The use of non-biological resources	1. Tidal electricity installations 2. Wind power installations 3. OTEC 4. Wave electricity 5. Marine current electricity ....
3. Tourist and leisure	...
4. Harbor and marine transportation	
5. Construction and engineering	
6. Marine R&D	
7. Environmental waste disposal	
8. Military and disaster use	
9. Indigenous people's traditional use of the sea	

Source: http://www.6law.idv.tw/6law/law2/%E9%99%84%E8%A1%A8%E4%B8%80%E4%B9%8B%E4%B8%80%E6%B5%B7%E5%9F%9F%E7%94%A8%E5%9C%B0%E5%AE%B9%E8%A8%B1%E4%BD%BF%E7%94%A8%E9%A0%85%E7%9B%AE%E5%8F%8A%E5%8D%80%E4%BD%8D%E8%A8%B1%E5%8F%AF%E4%BD%BF%E7%94%A8%E7%B4%B0%E7%9B%AE%E8%A1%A8.pdt.

In order to resolve the conflict of use (*fishery versus other purposes*, such as pipelines, data cables, dumping sites, national parks, natural conservation zones, cruise zones, etc.), the "Ordinance on the OWF Planning Sites" plays a potential role.

2. The application does not infringe upon the land utilization or environmental protection plan established by the central, municipal, or county (city) government based on the central or local autonomous laws and regulations.
3. Appropriate plans have been made for protection of the environment, conservation of nature, and prevention of disasters.
4. The application can match the water supply, nearby traffic facilities, drainage system, electric power, telecommunication, rubbish disposal, and other public facilities, equipment, and services.
5. The certificates of the rights of the lands and buildings in the area to be developed have been acquired.

74. Ordinance on Regulating the Use of Non-Urban Land 非都市土地使用管制規則http://law.moj.gov.tw/LawClass/LawContent.aspx?PCODE=D0060013.

Yet, the recent Three National Land Legislations would bring dramatic change to the long-existing regime. For instance, the Spatial Planning Act of 2015 provides a special role for "marine resource zones."[75] The Coastal Zone Management Act is also related to the development of OWF,[76] particularly the Coastal Conservation Plan.[77] The Wetland Conservation Act of 2013[78] is also relevant. New changes brought by these three laws are expected.

[5] International Law

Even though Taiwan is not a member of the *United Nations Convention on the Law of the Sea* (UNCLOS), it still promulgated the Act on Exclusive Economic Zone and the Continental Shelf of the Republic of China of 1998.[79]

The Republic of China shall enjoy and exercise sovereign rights of utilizing the energy stemming from the water, currents, winds, or other activities.[80] Moreover, the use of wind should be subject to permission.[81] Use without permission is subject to administrative penalty.[82]

[6] Public Participation Scheme

There is no specific public participation rule for OWF development. Yet, as EIA is mandatory during the application process, the OWF developers have to conduct a public hearing.[83] In addition, the aforementioned SEA could also be involved in a public hearing. Finally, there is also a special team under the National Energy Technology Research Program that oversees public communication and participation issues. For instance, recently, this team has been working diligently to deal with a fishery protest.

---

75. Articles 3, 4, 6, 7, 20, and 21 of the Spatial Planning Act of 2015.
76. http://law.moj.gov.tw/Eng//LawClass/LawContent.aspx?pcode = D0070222.
77. Article 10 of the Coastal Zone Management Act.
78. http://law.moj.gov.tw/Eng//LawClass/LawContent.aspx?pcode = D0070209.
79. Ministry of Justice, 1998. The Law on the Exclusive Economic Zone and the Continental Shelf of the Republic of China of 1998. Available at: http://law.moj.gov.tw/Eng/LawClass/LawContent.aspx?PCODE = A0000010 (accessed on August 15, 2016).
80. Article 5 of the Act on Exclusive Economic Zone and the Continental Shelf of the Republic of China of 1998.
81. Article 7 of the Act on Exclusive Economic Zone and the Continental Shelf of the Republic of China of 1998: "For utilizing energy from the water, currents, winds, or other activities in the exclusive economic zone of the Republic of China, a permission from the Government of the Republic of China shall be required. The related permission regulations shall be decided by the Executive Yuan."
82. Article 21 of the Act on Exclusive Economic Zone and the Continental Shelf of the Republic of China of 1998: "Whoever produces energy from the water, currents, winds, or other activities in the exclusive economic zone or on the continental shelf of the Republic of China without obtaining a permission from the Government of the Republic of China shall be punished with a fine of between two hundred thousand and one million New Taiwan Dollars, and the related equipment may be confiscated."
83. Article 12 of the EIA Act.

[7] Grid

The rules on laying an undersea electricity grid are governed by the Regulations of Permission on Delineation of Course for Laying, Maintaining, or Modifying Submarine Cables or Pipelines on the Continental Shelf of the Republic of China.[84] First, developers of cables or pipelines shall file an application in writing for a *course survey* to the MOIA.[85] After completing the course survey, developers shall draft a course plan and apply to the MOIA for course delineation permission.[86] Finally, the related fee is further defined in Fee-charging Standards of Permission for Laying, Maintaining, or Modifying Submarine Cables or Pipelines on the Continental Shelf of the Republic of China.[87]

Regarding the grid expansion and connection rules mentioned above, it is worth mentioning that to date there are no grid incentives, such as low-interest loans, to facilitate the development. This may cause an extra burden for "private" OWF developers but not for Tai-Power. As Tai-Power is a state-owned electricity company, they experience fewer problems obtaining sufficient money to develop the grid for their own OWF. Therefore, Tai-Power's double roles of OWF developer and grid owner may provide some advantage to developing an undersea grid cable.

[8] Other Concerns and the Legal Regime

Other concerns, such as national security-related regulations, flight safety, and shipment safety, are mainly dealt with by the aforementioned zonal regime. The environmental concerns, such as habitat and species protection and marine pollution, are mainly handled by the EIA or SEA as well as the zonal regime.

The relatively tough issues involve conflicting use with fishery zones. Fishermen usually protest against such development. With a view to resolving this, the public community team, under the National Energy Technology Research Program, has already conducted several communication events with the local fishermen. How to design a proper scheme to settle the dispute between OWF developers and the fishermen is also a key research topic.[88]

With a view toward gaining local support, a certain extent of the compensation regime is in place. For Tai-Power, there is an Electricity Infrastructure Development Fund Decree in place.[89] OWF has been listed as a compensation object under Article 6 of the decree. The compensation amount is NTD 300,000/MW. The site's

84. http://law.moj.gov.tw/Eng/LawClass/LawAll.aspx?PCode = K0060041.
85. Article 4 of the Regulations of Permission on Delineation of Course for Laying, Maintaining, or Modifying Submarine Cables or Pipelines on the Continental Shelf of the Republic of China.
86. Article 6 of the Regulations of Permission on Delineation of Course for Laying, Maintaining, or Modifying Submarine Cables or Pipelines on the Continental Shelf of the Republic of China.
87. http://law.moj.gov.tw/Eng/LawClass/LawContent.aspx?PCODE = D0060121.
88. www.ntpu.edu.tw/admin/a14/org/a14-1/files/announce/2015070791241.pdf.
89. Electricity Infrastructure Development Fund Decree (台灣電力股份有限公司促進電力發展營運協助金執行要點), http://www.taipower.com.tw/UpFile/RulesItemFile/enforce_keypoint960101.pdf.

village/district government is given 50%, while the other 50% is given to the county or municipality of the site.[90]

Yet, for private developers, there is no fixed amount of compensation like that of Tai-Power. It all depends on their communication or negotiation skill. For instance, recently there has been a dispute between Swancor and local fishermen on the amount of compensation for potential fishery loss after the installation of OWF. Swancor planned to pay NTD 350 million compensation for construction of thirty-two OWF turbines. Yet, the local fishery opposes such a proposal.[91] Thus, there is a discussion of whether to legalize or standardize the fishery-sector compensation formula under the Fisheries Act.[92]

§10.04 CHALLENGES AND SOLUTIONS

[A] Institutional Design: Challenges and Solutions

[1] Challenges

We found Taiwan's institutional OWF design to be fragmented horizontally and vertically. The lack of integration at the early stage led to only very abstract policy planning at the same. Yet, it seems the integrated style institutions, such as the New Energy Promotion Commission, the Executive Yuan (since 2009), may be very helpful in facilitating and specifying the development of OWF policy and program.

Yet, when it comes to the implementation, it seems the multi-authority model still causes uncertainty and is bothersome for OWF developers. The 1,000 Wind Power Projects Office began to tackle this issue in 2012; the development of the proposed four-to-six wind projects has shown promising progress. Still, some limitations seem to persist under this informal institution design.

[2] Solutions

The establishment of a "one-stop shop" or central authority to avoid the multi-authority pitfall is always the best way out.[93] It is recommended that perhaps the development of the OWF could follow the regime of major investment projects in the urban area. The RE Act of 2009 could be revised to introduce legal provisions similar to Article 29-2 of the Urban Planning Act. A joint examination meeting or review meeting could be formed to simplify a multi-authority review, as stated in the following:

> Parallel operations may be adopted for major development investment cases that involve formulation or modification of urban plans and environmental impact

90. Article 25 of Electricity Infrastructure Development Fund Decree.
91. http://news.ltn.com.tw/news/society/breakingnews/1326791.
92. Fisheries Act, http://law.moj.gov.tw/Eng/LawClass/LawContent.aspx?PCODE=M0050001.
93. Gao, *supra* note 19, at 14.

assessment, water conservation, and maintenance required by law. If necessary, the competent authority for the urban development in concern may convene *joint examine meetings* to make the decisions.

The possible Plan B could be related to the Organizational Act of the Executive Yuan of 2010. During that reform, the MOEA was transformed into an authority similar to the UK's Department of Energy and Climate Change (DECC).[94] Perhaps such a situation would facilitate the development of OWF projects.

Finally, reinforcement at the top level could also be vital. Since 2009, high-level commissions have been in place, including the New Energy Promotion Commission, the Executive Yuan (since 2009), the Green Energy and Low-Carbon Promotion Commission (since 2014), the Green Energy and Low-Carbon Promotion Commission (since 2014), and, this year, after the commission of the new government, an Office for Energy Conservation and Emission Reduction (since 2016) that was created to intensify the coordination work at the cabinet office. Such development may be very helpful in identifying coordination problems and facilitating OWF projects.

[B] Incentives: Challenges and Solutions

[1] Challenges

As noted by one article, there are several weaknesses in Taiwan's incentive scheme.[95]

First, the main discussion focuses only on the "main" scheme, i.e., the tariff rate, and thus seems to overlook the importance of "supplementary" schemes such as loans, the interest from loans, and tax credits, particularly regarding the grid connection rule.

Second, it is doubtful that the FIT rate will be sufficient to encourage the further development of OWFs.

Third, an insufficient supplementary scheme may cast a shadow over the development of OWFs. According to recent news, developers seem to be facing a problem with loans and guarantees - many bank loans have been pending for a long

94. Article 3 of the Organizational Act of the Executive Yuan of 2010: "The Executive Yuan establishes the Ministries as follows:

(1) Ministry of the Interior;
(2) Ministry of Foreign Affairs;
(3) Ministry of National Defense;
(4) Ministry of Finance;
(5) Ministry of Education;
(6) Ministry of Justice;
(7) Ministry of Economic and Energy Affairs;
(8) Ministry of Transportation and Construction;
(9) Ministry of Labor;
(10) Ministry of Agriculture;
(11) Ministry of Health and Welfare;
(12) Ministry of Environment and Natural Resources;
(13) Ministry of Culture; and
(14) Ministry of Science and Technology."

95. Gao, *supra* note 19, at 9.

time.[96] Moreover, granting tax credits for excise duties, rather than more incentives such as accelerated depreciation, may not be very helpful in providing impetus.

Finally, the problematic and unclear grid connection rules may cast a further shadow on the OWF project because this part of the cost may not be taken into account when determining the FIT. Thus, an OWF wind developer may face cost recovery issues and loan issues due to such an expensive connection. The current Article 8 of the RE Act should be revised to introduce a "super-shallow connection" for OWFs. Currently, Article 8[2] of the RE Act states that "the said parallel connection with proper technology is limited to enterprises with economical and reasonable cost burdens; other than the existing circuits, the cost for power grid enhancement will be shared among both the electricity enterprises and renewable energy power facility installers." Article 8(5) of the Act continues: "The installer shall install and maintain the circuits connecting renewable energy power facilities and power grids; the electricity enterprises with which its power facilities are paralleled to the installer's facilities shall provide assistance when necessary; the renewable energy power generation facility installers shall bear the costs incurred."

[2] Solutions

The current main scheme FIT has already been enhanced several times, and a relatively flexible regime (for OWF developers to choose a fixed rate for twenty years or a front-end tariff for two ten-year periods) was introduced. This may be sufficient for OWF development. Yet, when taking into account the unforeseen grid costs and the potential delayed construction costs (from the lessons of Germany), it is unclear whether such a FIT rate is sufficient. Thus, in order to enhance investment security, there seems to be a need to introduce the super-shallow connection model.

For the supplementary scheme, a low-interest loan and guarantee regime managed by a public bank needs to be reconsidered. Perhaps there is a need to add a one-use purpose in the spending provisions of the Renewable Energy R&D Fund under Article 7(3) of the RE Act.[97] Without this, the OWF may again be financed by foreign banks that encourage the use of their own OWF turbines. In addition, it should be determined whether the tax exemption provision of Article 16 can be expanded to other taxes. Of course, the investment subsidy and demonstration subsidy could be used in a flexible way to respond to the insufficiency of FIT, taking into account all relevant costs, such as grid costs.

---

96. http://www.businesstoday.com.tw/article-content-92751-113067?page=3.
97. Article 7(3) of the RE Act states: "Regulations governing the collection, procedure, deadline, and other relevant matters regarding the fund referred to in paragraph 1 shall be prescribed by the central competent authority.

The aforesaid fund in paragraph 1 shall be used for the following purposes:

1. to subsidize renewable energy electricity fees.
2. to subsidize renewable energy facilities.
3. to subsidize demonstration and promotion of the use of renewable energy.
4. for other purposes related to the development of renewable energy approved by the central competent authority."

[C] Regulations: Challenges and Solutions

[1] Challenges

The long-existing electricity-generation licensing scheme designed for non-liberalization of the market may not fit the new dynamic market situation. The three-stage bothersome licensing process causes many challenges for the OWF developers.

The current EIA and SEA regime seems to overburden the OWF developers as well. The unique EIA veto and permit nature in Taiwan also poses a threat to the development of relatively environmental-friendly low-carbon technology. In addition, the zero threshold for OWF is problematic. The most serious issue is that the review and high scrutiny culture of the EIA seems to affect the original consultation nature of the SEA, as well. This makes the original purpose of the SEA, which is to facilitate OWF development, become an additional barrier instead.

The new planning laws may also add extra burden to the OWF projects. In addition, it may affect the existing OWF projects. Moreover, the extra legal requirement means more red tape must be navigated, causing need for additional coordination of the administrative procedures.

Finally, extra concerns may also pose a threat. For instance, the protest against the OWF and the compensation scheme as a tool to deal with such protests may convey the wrong signal and generate a vicious cycle in OWF project development.

[2] Solutions

I propose certain ways out.

(1) License regime:
- Installation regulations and electricity-generation business regulations should be separated to simplify the process.
- Consider allowing the OWF back under the jurisdiction of the RE Act and thereby subject to its relatively fast-track review process, instead of the Electricity Business Act.
- The Taiwan government should promulgate a special Registration Ordinance for Offshore Wind Power in Taiwan (the Chinese government has enacted a special ordinance in order to resolve the complex license issues affecting OWFs).

(2) EIA:
- Like the chance taken recently in reforming the EIA process,[98] the EIA veto should be reformed into the normal advisory system.

98. Administration to Review EIA system, Taipei Times, June 28, 2016; EPA Touts Review of EIA Act, Taipei Times, June 21, 2016.

- Article 29 of the EIA Ordinance should take into account the needs of OWF. An EIA exemption for small-scale, pilot, or demonstration OWFs should be considered. There should be an MW screening threshold and installation numbers for facilitating the screening process of the EIA. When determining the threshold, the relatively rigid nature of the EIA in Taiwan should be taken into account.

(3) SEA:
- The Ordinance on Mandatory SEA Policy and Program should add the item of the "Offshore Wind Power Plan/Program" to the category of "Land Use Policy."
- SEA should be returned to its consultation nature, instead of setting an extra condition for the project EIA.

(4) Planning law:
- There is a need to simplify the new permit and procedure that resulted from the new "Three National Land Acts."
- Follow the Urban Planning Act and create a joint review committee.

(5) Compensation and public communication:
- The potential side effects of the legalization of the compensation scheme to local people and fishery sectors should be evaluated.

§10.05 CONCLUSION

In order to realize a nuclear-free homeland in such an isolated island electricity system and a highly manufacturing-based country like Taiwan, while taking into account the tough job of tackling international emission reduction obligations, Taiwan has limited energy choices for the future. The development of low carbon energy requires a total solution. In the past few years, a relatively comprehensive framework has been in place to push forward the development of PV and onshore wind power. Yet, in spite of long-term efforts, such as the First Phase Offshore Wind Program of 2007, the OWF finally deployed two turbines last year. This also can show how difficult it is to apply and introduce OWF in Taiwan.

What could be the main reason behind such a delay? Perhaps one of the reasons could be tough geological locations for installing OWF. Yet, if this is such the case, why are so many foreign companies interested in Taiwan's surrounding wind resources? Apparently, the insufficient promotion framework may be to blame. It began with a qualitative policy without a quantitative target and then became a scheme too inclined toward the FIT rate while ignoring the supplementary incentive schemes. The recent development seems to be overkill and to overcorrect the current situation, and too great of a regulatory burden could kill the future of OWF in Taiwan.

During this process, Taiwan has tried diligently to learn the legal regime of other countries. Why, then, does it fail? This may be caused by incomplete learning from other countries' experiences. Examples include the FIT for OWF did not take into account the grid cost at the early stage, the grid cost-sharing rules do not take into account the needs of the OWF, the unique EIA threshold of zero for OWF, the peculiar operation of SEA, etc. All of these partial transpositions of the UK and Germany regimes into Taiwan lead to new questions and troubles.

How do we resolve such an impasse? I do not have an answer at this time, but it seems that the law has been insufficiently revised to reflect the needs of OWF. Thus, I recommend more legal revisions, as they could be the best resolution. Revision is necessary that better coordinates legislation. It is hoped that under the leadership of the newly established Office for Energy Conservation and Emission Reduction, a more comprehensive solution could be proposed in the future.

Index

T

ENERGY AND ENVIRONMENTAL LAW & POLICY SERIES

1. Stephen J. Turner, *A Substantive Environmental Right: An Examination of the Legal Obligations of Decision-makers towards the Environment*, 2009 (ISBN 978-90-411-2815-7).
2. Helle Tegner Anker, Birgitte Egelund Olsen & Anita Rønne (eds), *Legal Systems and Wind Energy: A Comparative Perspective*, 2009 (ISBN 978-90-411-2831-7).
3. David Langlet, *Prior Informed Consent and Hazardous Trade: Regulating Trade in Hazardous Goods at the Intersection of Sovereignty, Free Trade and Environmental Protection*, 2009 (ISBN 978-90-411-2821-8).
4. Louis J. Kotzé and Alexander R. Paterson (eds), *The Role of the Judiciary in Environmental Governance: Comparative Perspectives*, 2009 (ISBN 978-90-411-2708-2).
5. Tuula Honkonen, *The Common but Differentiated Responsibility Principle in Multilateral Environmental Agreement's: Regulatory and Policy Aspects*, 2009 (ISBN 978-90-411-3153-9).
6. Barbara Pozzo (ed.), *The Implementation of the Seveso Directives in an Enlarged Europe: A Look into the Past and a challenge for the Future*, 2009 (ISBN 978-90-411-2854-6).
7. Henrik M. Inadomi, *Independent Power Projects in Developing Countries: Legal Investment Protection and Consequences for Development*, 2010 (ISBN 978-90-411-3178-2).
8. Nahid Islam, *The Law of Non-Navigational Uses of International Watercourses: Options for Regional Regime-Building in Asia*, 2010 (ISBN 978-90-411-3196-6).
9. Yasuhiro Shigeta, *International Judicial Control of Environmental Protection: Standard Setting, Compliance Control and the Development of International Environmental Law by the International Judiciary*, 2010 (ISBN 978-90-411-3151-5).
10. Katleen Janssen, *The Availability of Spatial and Environmental Data in the European Union: At the Crossroads between Public and Economic Interests*, 2010 (ISBN 978-90-411-3287-1).
11. Henrik Bjørnebye, *Investing in EU Energy Security: Exploring the Regulatory Approach to Tomorrow's Electricity Production*, 2010 (ISBN 978-90-411-3118-8).
12. Véronique Bruggeman, *Compensating catastrophe victims: A Comparative Law and Economics Approach*, 2010 (ISBN 978-90-411-3263-5).
13. Michael G. Faure, Han Lixin & Shan Hongjun, *Maritime Pollution Liability and Policy: China, Europe and the US*, 2010 (ISBN 978-90-411-2869-0).

14. Anton Ming-Zhi Gao, *Regulating Gas Liberalization: A Comparative Study on Unbundling and Open Access Regimes in the US, Europe, Japan, South Korea and Taiwan*, 2010 (ISBN 978-90-411-3347-2).
15. Mustafa Erkan, *International Energy Investment Law: Stability through Contractual Clauses*, 2011 (ISBN 978-90-411-3411-0).
16. Levente Borzsa´k, *The Impact of Environmental Concerns on the Public Enforcement Mechanism under EU law: Environmental protection in the 25th hour*, 2011 (ISBN 978-90-411-3408-0).
17. Tarcísio Hardman Reis, *Compensation for Environmental Damages under International Law: The Role of the International Judge*, 2011 (ISBN 978-90-411-3437-0).
18. Kim Talus, *Vertical Natural Gas Transportation Capacity, Upstream Commodity Contracts and EU Competition Law*, 2011 (ISBN 978-90-411-3407-3).
19. WangHui, *Civil Liability for Marine Oil Pollution Damage: A Comparative and Economic Study of the International, US and Chinese Compensation Regime*, 2011 (ISBN 978-90-411-3672-5).
20. Chowdhury Ishrak Ahmed Siddiky, *Cross-Border Pipeline Arrangements: What Would a Single Regulatory Framework Look Like?*, 2012 (ISBN 978-90-411-3844-6).
21. Rozeta Karova, *Liberalization of Electricity Markets and Public Service Obligations in the Energy Community*, 2012 (ISBN 978-90-411-3849-1).
22. Sandra Cassotta, *Environmental Damage and Liability Problems in a Multilevel Context: The Case of the Environmental Liability Directive*, 2012 (ISBN 978-90-411-3830-9).
23. Mark Wilde, *Civil Liability for Environmental Damage: Comparative Analysis of Law and Policy in Europe and US*, 2013 (ISBN 978-90-411-3233-8).
24. Bernard Taverne, *Petroleum, Industry and Governments: A Study of the Involvement of Industry and Governments in Exploring for and Producing Petroleum*, 2013 (ISBN 978-90-411-4563-5).
25. Anton Ming-Zhi Gao & Chien Te Fan (eds), *Legal Issues of Renewable Energy in the Asia Region: Recent Developments in a Post-Fukushima and Post-Kyoto Protocol Era*, 2014 (ISBN 978-90-411-4856-8).
26. Sabina Manea, *The Instrumentalization of Property: Legal Interests in the EU Emissions Trading System*, 2014 (ISBN 978-90-411-5420-0).
27. Joseph A. Tolorunse, *Protection of Property Rights in Discovered Petroleum Reservoirs*, 2014 (ISBN 978-90-411-5604-4).
28. Katelijn Van Hende, *Offshore Wind in the European Union: Towards Integrated Management of Our Marine Waters*, 2015 (ISBN 978-90-411-5613-6).
29. Ken'ichi Matsumoto & Anton Ming-Zhi Gao (eds), *Economic Instruments to Combat Climate Change in Asian Countries*, 2015 (ISBN 978-90-411-5408-8).

30. Eduardo Pereira (ed.), *Joint Operating Agreements: Challenges and Concerns from Civil Law Jurisdictions*, 2015 (ISBN 978-90-411-5934-2).
31. Ying Shen, *China's Way to Carbon Emissions Reduction: The Choice of Regulatory Instruments and Its Legal Challenges*, 2015 (ISBN 978-90-411-6049-2).
32. Roy Andrew Partain, *Environmental Hazards from Offshore Methane Hydrate Operations: Civil Liability and Regulations for Efficient Governance*, 2017 (ISBN 978-90-411-8730-7).
33. Anton Ming-Zhi Gao & Chien-Te Fan (eds), *The Development of a Comprehensive Legal Framework for the Promotion of Offshore Wind Power: The Lessons from Europe and Pacific Asia*, 2017 (ISBN 978-90-411-8397-2).